CONTENTS

FOREWORD

Mining and environment — the two do not seem to go together. Indeed, they seem almost antithetical. Whether one reads about small-scale gold mining in the Amazon or huge coal mines in North America, whether simple sand and gravel pits or complex metallurgical operations, the legacy of the mining industry appears to be destruction of land and pollution of air and water. Actually, of course, the situation is much more complex. True, mining always involves disruption of the environment, either at the surface with open-pit mines or underground with deep mines, and in most cases the mineral being sought makes up only a small part of the material that must be moved, with the result that vast quantities of waste must be handled. True too, for many years and in most parts of the world (the North no less than the South), mining was carried on with little regard for environmental protection — or for the health and safety of miners or for the culture and well-being of local communities. However, the picture of mining firms operating with little regard for nature or native is no longer accurate. Under some conditions, and in some corporations, and in some countries, protection of the environment, of miners, and of nearby communities has become nearly as much a concern as putting a rock in the box.

International Development Research Centre (IDRC) funding is based on the principle that solutions to problems in developing countries can only be found through research based in those countries. From this perspective, it is not so much the record of past destruction of the natural environment that is of interest, but the dynamics of a new business environment in which corporate decisions and government legislation work in tandem to avoid damages that are avoidable and to mitigate those that are not. Of course, there remain as many cases where the old conditions persist, and it is equally of interest to learn where and why this new business environment has not appeared.

The early conclusions of the Mining and Environment Research Network (MERN[1]) were striking: mining firms that are efficient in their main activity of extracting minerals from the Earth are also best at protecting the environment

---

[1] Originally based at the Science Policy Research Unit, University of Sussex, Unite Kingdom; now based at the International Centre for the Environment, University of Bath, United Kingdom.

while doing so, a conclusion that suggests that, under the right conditions, the economics–environment trade-off is not so sharp as once thought. The further conclusion that environmental results are partially independent of the strength of mining legislation suggests the need for governments to take a more sophisticated approach to mining-environment policy. Corporate attitudes are changing; government policies are changing; civil society is changing: and the business environment that brings them all together is changing. The need for further research is almost self-evident.

IDRC's interest in mining in developing countries predates today's recognition of environmental values and its focus on mining-environment policy. During the 1980s, research projects funded by IDRC focused mainly on science and technology policy for mining or on measures to improve efficiency. With the partial exception of a couple of projects that investigated health conditions in Bolivian mines (in particular, the effects of living and working at high altitudes), environment was very much secondary. An explicit environmental project related to mining does not appear until 1991, and somewhat ironically the first such project was the Bolivian component in the initial phase of MERN, as described in Chapter 8 in this book. Other projects looked at the effects of mercury from gold mining. Closely related projects also began to be funded, including some that focused less on environmental problems per se than on conflicts that stemmed from the power of the mining industry to usurp what had been common-property resources. A good example was the dispute over water that occurred between Southern Peru Copper Company and the community of Ilo, just north of Peru's border with Chile. (That research project was undertaken by a community group called LABOR, which then argued its case successfully before the International Water Tribunal in The Hague.) Another line of research that was initiated about the same time involved the effects of macroeconomic conditions and policies in various Latin America countries on the linkage between environmental degradation and mineral operations.

If the analytical focus of mining research funded by IDRC changed from technical efficiency to environmental protection, the geographic focus did not. With the exception of a collection of projects that focused on artisanal mining (mainly for gems) in Africa and Asia, the projects have almost all involved Latin America. This emphasis is not surprising: IDRC's program in South America focuses on the Andean countries, and this region is, perhaps more than any other in the world, dependent on mining for economic health. This focus is likely to continue. As this book appears, IDRC is conducting an inventory of research and researchers on mining and environment in the continent. From this, it is hoped that a long-term strategy for a coordinated program of research, probably focusing on ecosystem health, will emerge. (Ecosystem health is a new approach that links the effects on human health that stem from adverse anthropogenic changes to the natural environment.) The objective would be to determine what changes in government or corporate policies and what forms of community involvement in decision making would do most to protect local and regional ecosystems and, therefore, human health.

For the time being, however, what is needed is analytically sound documentation of the extent and the effects of recent changes in the business environment, as reflected in corporate behaviour and government policy. This is done very effectively, and for a wide range of corporations and conditions, in this first book from MERN.

David B. Brooks
Chief Scientist
International Development Research Centre
Ottawa, Canada

Acknowledgments

The editor would like to extend particular thanks to Lisa Eisen, whose tireless and diligent editorial work was vital in preparing these chapters for publication. Special thanks are also due to Yvette Haine and Gavin Bridge for their help with the original manuscript. This book reflects the insight and hard work of many researchers and affiliates of the Mining and Environment Research Network (MERN) whose comments during MERN's annual research meetings contributed to the research contained within these pages. Finally, sincere thanks go to the staff of the International Development Research Centre, particularly to David Brooks, Bill Carman, Brent Herbert Copley, Amitav Rath, and Ann Whyte, for their hard work, from project development to completion.

Introduction — MERN: Toward an Analysis of the Public Policy–Corporate Strategy Interface

This book describes a process of building research capacity in the area of mining and the environment. This process began in the mid-1980s in Latin America, when a number of policy researchers — working together within International Development Research Centre (IDRC)-supported projects on issues of competitiveness, production efficiency, and technological change — began observing a noticeable association between production inefficiencies and environmental damage.

This was at a time when countries such as Bolivia, Brazil, Chile, and Peru either had a poorly developed or nonexistent environmental regulatory regime or lacked the institutional capacity to implement environmental policy. An early observation, examined in detail in this book, was that environmental regulation seemed to fail as the prime determinant of good environmental practice. Environmental performance of firms seemed to differ as much within one regulatory regime as between different regulatory regimes. These observations suggested the need for empirical research at the firm and plant levels, both to describe the environmental practices observed and to check these against the types of operations, vintages of technology, and competitive situations of the minerals sectors of the Americas. Our rationale was to learn how to improve public policies and corporate strategies for the environment.

This research was unfolding in two interrelated contexts. First, many mineral-producing countries were entering a period of economic liberalization, with privatization of previously state-owned mining companies and investment regimes opening up to attract international mining firms. Second, an environmental imperative was fast emerging, with increasingly stringent environmental regulations; growing voice-of-society concerns; environmental conditions often being attached to credit and insurance for new or expanding mining operations; and environmental pressures appearing throughout the supply chain, as industry was increasingly demanding that its suppliers meet new environmental standards. It is also important to recognize that this research was conducted in the early 1990s, when the importance of such diagnostic work was first recognized and we urgently needed descriptive analysis to be able to draw policy lessons.

A collaborative framework was devised to undertake this research, and proposals for its funding were submitted to a number of key agencies with portfolios in environment and development. Seed funding from the Organisation for Economic Co-operation and Development (OECD) Development Centre and the Overseas Development Administration made it possible, in 1991, to establish the Mining and Environment Research Network (MERN), a collaborative project involving researchers from the following institutions: the University of São Paulo and the Centre for Mineral Technology (Centro de Tecnologia Mineral) in Brazil; the Institute for Research on Public Health (Instituto de Salud Popular) and the Catholic University of Peru (Pontificia Universidad de la Católica del Perú) in Lima, Peru; the Centre for Studies in Mining and Development (Centro de Estudios Mineria y Desarrollo) in La Paz, Bolivia; and the Centre for Studies of Copper and Mining (Centro de Estudios del Cobre y de la Mineria) in Santiago, Chile. The collaborative research project developed and won the prestigious John D. and Catherine T. MacArthur Foundation Collaborative Studies competition in 1991. Together with complementary funding for the Bolivia project from IDRC, this launched the first phase of MERN, which is the subject of this book. MERN fast expanded to include a range of different types of interdisciplinary centres of excellence in mineral-producing developing and industrialized countries. The current list of members is summarized by institution and country in Table 1.

From the outset, MERN aimed to provide research analysis to inform environmental public policy and to help mining companies achieve environmental compliance and improve competitiveness in the context of growing environmental regulation and technological innovation (see Box 1). The international collaborative research program first set out to examine the relationship between environmental regulation, technical change, and competitiveness in the nonferrous-minerals industry. In particular, it investigated how the process of technological innovation and organizational change could be harnessed to prevent environmental degradation while enhancing productivity and sustainability.

The rationale for the international focus of the research effort, in the developing and industrialized countries, was the need to learn from the experience of more competitive and environmentally proactive firms while focusing on the challenges of achieving environmental best practice in each participating country. The reason for adopting a network approach, as opposed to working independently, was to build up an international pool of interdisciplinary research competence. In other words, research capacity-building was a goal of the network from the outset.

Table 1. MERN: a collaborative research network.

Country	Institutions
United Kingdom	International Centre for the Environment at the University of Bath University of Sussex University of Surrey University of Dundee Camborne School of Mines Royal School of Mines
Argentina	Gerencia Ambiental CIS University of Buenos Aires
Australia	University of South Australia University of Murdoch
Bolivia	Centro de Estudios Mineria y Desarrollo
Brazil	University of São Paulo Centro de Tecnologia Mineral
Bulgaria	Geological Institute of Bulgarian Academy of Sciences
Canada	Centre for Resource Studies at Queen's University at Kingston
Chile	Centro Estudios del Cobre y de la Mineria Catholic University of Chile
China	Eco-Environmental Research Center Academia Sinica
Colombia	Universidad Pontificia Bolivariana Instituto de Estudios Regionales
Ethiopia	Mineral Resources for Exploration and Development Addis Ababa University
France	École des mines
Germany	Projekt Consult Oeko Institute for Applied Ecology Universität Gesamthochscule
Ghana	Minerals Commission Institute of Management and Public Administration Environmental Protection Agency
Hungary	Minerals Commission
India	Tata Energy Research Institute National Institute of Small Mines Central Mining Research Institute
Italy	Universita degli Studi di Cagliari
Japan	Hokkaido University
Malaysia	National University of Malaysia
Mozambique	Eduardo Mondlane University
Netherlands	University of Amsterdam
Norway	University of Oslo
Pakistan	University of the Punjab

(concluded)

Table 1 concluded.

Country	Institutions
Papua New Guinea	Department of Environment and Conservation
Peru	Instituto de Salud Popular Pontificia Universidad de la Católica del Perú
Poland	University of Mining and Metallurgy
Republic of Guinea	Ministry of Mines and Geology
South Africa	Mitsubishi Electric Personal Computer Division Mintek University of Cape Town
Sweden	Raw Materials Group University of Lund
Tanzania	Ministry of Energy and Minerals
Thailand	Thailand Development Research Institute Prince of Songkla University
United States	Colorado School of Mines Massachusetts Institute of Technology US Bureau of Mines East–West Center, Hawaii
Zambia	Mining Sector Co-ordinating Unit South Africa Development Community
Zimbabwe	University of Zimbabwe Institute of Mining Research John Hollaway Associates

Note: MERN, Mining and Environment Research Network.

A number of environmental concerns made us decide to work collaboratively toward our research objectives. Our early research profiles were all in the area of minerals policy and technological change in the mining sector and in the countries where we worked; many of us had done field work in the Andean region. We were all increasingly observing environmental damage associated with minerals extraction or processing activities: Figure 1 shows the relationships between the mining process, its waste products, and the hazards they present. This relates to environmental impacts affecting the three environmental media of land, water, and air, as well as the effects on local communities. Although in each country and at each mine site, environmental damage differed, some common explanatory factors seemed to emerge. Some of these, we immediately recognized, flew in the face of conventional wisdom. Conventional wisdom maintains the pollution-haven thesis that suggests international firms locate their production activities where they can most easily externalize the environmental-damage costs of their

Box 1

The Mining and Environment Research Network

MERN is an international collaborative research program involving centres of excellence in the major minerals-producing countries of the world. The program was established in 1991, with the aim of helping mining companies to achieve environmental compliance and improve competitiveness in the context of growing environmental regulation and technological innovation.

Our current research examines the relationship between environmental regulation, technical change, and competitiveness in the nonferrous-minerals industry. We investigate how the processes of technological innovation and organizational change can be harnessed to prevent environmental degradation while enhancing productivity and sustainability. The liberalization of investment regimes worldwide, combined with growing environmental regulation and more frequent requirements for an environmental-impact analysis as a precondition for loans, means that objective and well-documented policy analyses are urgently needed to support decision-making in industry, donor agencies, government, and nongovernmental organizations. This program of collaborative research aims to facilitate the global diffusion of such policy analyses and contribute to building international research competence in this area.

Taking this into account and building on our diagnostic research, the next phase of MERN research covers three interrelated themes:

- *Comparative analysis of environmental performance and its relationship with production efficiency* — MERN research has demonstrated that good environmental management in the firm is more closely related to production efficiency and capacity to innovate than to regulatory regime. Environmental degradation tends to be greatest in high-cost operations working with obsolete technology, limited capital, and inadequate human-resource management. Because these problems are characteristic of much of the minerals production of developing countries, they are a special, but not exclusive, focus of MERN research. A major area of empirical research is an international benchmarking exercise to investigate environmental performance.

- *Analysis of international environmental regulations and the definition of improved policy options* — Building on an international comparative analysis of the effectiveness of current environmental regulations, researchers are investigating a range of policy approaches to achieve sustained and competitive improvements in environmental management and to achieve pollution prevention, as opposed to pollution treatment. The research will make an original contribution by evaluating the potential of technology transfer and training (particularly if governed by joint-venture agreements and are linked to credit conditionality) to accelerate the development and diffusion of improved environmental-management practices. Researchers are also analyzing the environmental implications of new trade policies and agreements, such as the General Agreement on Tariffs and Trade and the North American Free Trade Agreement.

- *Toward best practice: corporate trends in environmental management* — A preliminary conclusion of MERN research is that technical change, stimulated by the drive for improved competitiveness and the environmental imperative, is reducing both production and environmental costs, to the advantage of those companies that have the resources and capacity to innovate. Our current phase of research is intended to evaluate and compare trends in environmental best practice for nonferrous-minerals production in different socioeconomic and policy contexts, drawing out the lessons for both corporate strategy and government policy. This includes empirical research on planning for closure within the minerals industry.

(continued)

Box 1 concluded.

Past and current sponsors of MERN research and dissemination activities include the John D. and Catherine T. MacArthur Foundation; IDRC, Canada; the Overseas Development Administration, United Kingdom; the Economic and Social Research Council, United Kingdom; the OECD, Paris; the US Bureau of Mines; the United Nations Environment Programme, Paris; the Science, Technology, Energy, Environment and Natural Resources Division of the United Nations; Industry, Science and Technology Canada and Environment Canada; the Chinese State Science and Technology Commission; the Columbian Institute for the Development of Science and Industry (Instituto Columbiano para el Desarrollo de la Ciencia y Tecnologia), Colombia; the British Council; and a growing number of MERN Industry Club sponsors.

The output of MERN includes ongoing publication of research articles and reports, conference papers, books (including edited volumes of case studies), a biannual bulletin and briefing papers for sponsors, national workshops, and an annual international conference. As the MERN members develop research capability and define new areas of work and as demands on MERN's central resources increase, new funding is being sought. The benefits for MERN's sponsors include full access to MERN's central services and research findings (which include the results of detailed empirical studies in most of the major minerals-producing countries) and to a network of contacts, including interdisciplinary teams in well-placed centres of excellence.

For further details on membership, sponsorship, or research, please contact

Professor Alyson Warhurst
Director of the Mining and Environment Research Network
International Centre for the Environment
School of Management
University of Bath
Bath, UK BA2 7AY
Tel: +44 (0)1225 826156
Fax: +44 (0)1255 826157

---

production, that is, in developing countries where environmental regulations are either limited or poorly enforced. However, an early, common observation explored in these studies was that environmental damage was not evenly distributed within the minerals sector of each developing country studied but that seemed to vary according to a number of other factors, such as type of mineral; vintage of technology; stage of investment; stage of operation; level of integration; effectiveness of environmental regulation and its enforcement; and socioeconomic context (including poverty in local communities and work-force education and training). Most of all, environmental performance varied according to the firm's inherent technological dynamism.

These studies are therefore unique in refusing to accept a priori conventional wisdom and seeking to investigate actual environmental performance and its determinants at the mine site and across a range of site-specific factors.

The intellectual benefits of working together in a network and collaborating across a range of research and dissemination activities were greater than they would have been as the sum of the efforts of each institution working alone. Our

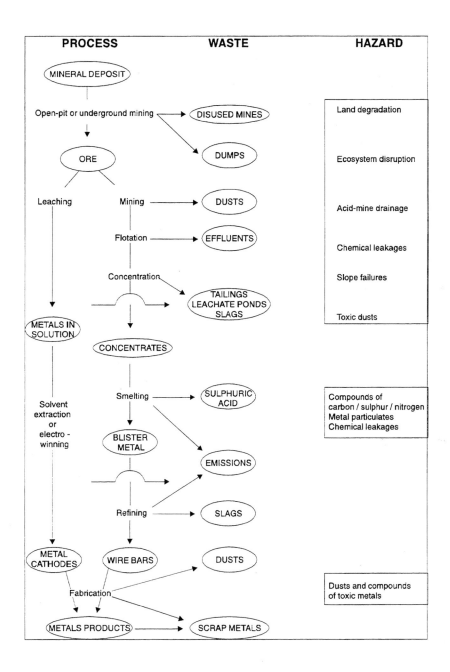

Figure 1. The mining process and the environment. Source: Warhurst (1991a).

networking activities included the coordination of our efforts through an electronic database and information system. As we grew from an initial 6 institutions to 56, luckily so did information technology grow in sophistication. We now do most of our networking activities via electronic communication. The network produces a bulletin twice annually, and this contains progress reports from each group, policy updates, industry news, professional articles, and a conference calendar. We recently completed Bulletin No. 10; collectively, the bulletins record MERN's growth over the years.

Once a year, MERN meets for a research workshop on a theme-by-theme basis. Workshop themes have included mining and sustainable development; pollution prevention, risk, and responsibility; planning for closure; and best practice in the management of mining's ecological impacts. Each workshop includes special sessions for research feedback and dissemination. Each of the chapters of this book went through such a peer-review process.

Finally, competence-building was absolutely paramount in this research process — these research studies were also undertaken as an education and training exercise. Two of the researchers who contributed chapters to this book, Maria Hanai and Fernando Loayza, were awarded PhDs for their work. All the researchers, including the editor, Alyson Warhurst, built on the network-research experience to develop their research programs. Each has now achieved either promotion within an existing role or a new position of decision-making responsibility relating to minerals development in their countries.

Organization

It remains to briefly review each chapter and describe the perspectives of the authors on these emerging themes in mining and the environment.

In Chapter 1, Alyson Warhurst develops a framework for the analysis of the case studies of environmental practices. Building on her training in the Earth sciences and in policy research, she describes the characteristics of dynamic firms that innovate to prevent pollution rather than reducing it after it has occurred. She analyzes these cases from the perspective of companies' strategies to externalize and internalize the environmental-damage costs of their production. Dynamic firms are those that not only internalize the environmental-damage costs of their production in response to the environmental imperative but also innovate to reduce their direct production and future abatement costs and so diminish the environmental–economic trade-off that conventional economic theory regards as the constraint on environmental progress. She then uses these findings to analyze changes in environmental-regulation approaches. A paradigm shift is occurring from command-and-control environmental regulations to ensure the polluter pays to environmental regulations governed by the principle that pollution prevention pays. However, Warhurst concludes that current policy mechanisms fail to promote technological and organizational change within firms to ensure pollution is prevented from the outset. She recommends broadening the range of regulatory mechanisms and making the technology-policy mechanisms and economic instruments needed to support them combine both regulation and promotion of industrial development. This new approach she terms *environmental innovation*.

In Chapter 2, Juanita Gana draws on her training in Chile as an engineer and her postgraduate training in the United States in minerals economics to examine the US experience of developing regulatory approaches in the minerals sector. She investigates waste disposal and the control of SO_2 emissions and concludes that even US environmental policy is still very much in the trial-and-error phase and that solutions to environmental threats are constrained by our lack of scientific knowledge about the ecosystem and the impacts of human activities like mining. Gana ends by drawing some lessons for Chile US from experience. She concludes that, particularly in a developing country, cost-effective policies are crucial and that site-specificity should be examined to ensure that regulations are relevant to the site-specific pollution hazards. She argues for the ecoregional administrative approach to economic policy and therefore to environmental regulation in Chile and for a case-by-case approach to negotiating site-specific environmental controls. Gana highlights Chile's potential to learn from the mistakes and failures of other regulatory regimes and to begin its efforts with more appropriate environmental-policy objectives.

In Chapter 3, Gustavo Lagos, an engineering specialist in the mining industry, and Patricio Velasco develop these ideas through an analysis of environmental policies and practices in Chilean mining. They trace the environmental impacts of mining in Chile back to colonial times but recognize that environmental problems became more acute during the 1980s because of the growth of mining. A progressive impoverishment of ore grades was leading to increasing volumes of metallic impurities in tailings from processing plants and in smelter-feed material. Public awareness in Chile and international concern about environmental impacts also grew during the 1980s. Lagos's earlier research identified SO_2 and particulate emissions from the country's six smelters as the key sources of pollution, followed closely by leakages from tailing dams and leaching operations, resulting in river and sea contamination.

Because of the lack of previous research in Chile on environmental pollution from mining and its policy, Lagos and Velasco's study mainly describes and identifies practices, trends, and policy issues. A major part of their analysis focuses on the very diverse environmental criteria adopted by different regional agencies and ministries and the diverse approaches adopted by the regulatory authorities to state, national private, and international operations. The authors report that several international mining firms adopted environmental practices in advance of legislated norms and institutional recommendations. But the state-owned companies face massive challenges in dealing with their sins of the past, in terms of accumulated environmental problems, combined with other factors such as the state companies' history, culture, and resource constraints.

The authors also report that consensus is more commonly obtained in other spheres of Chile's political, economic, and social development, such as in industrial development and the role of the private sector, than in environmental policy. Apparently, some people believe that sacrifices in the quality of the environment are needed to achieve fast economic development. Lagos and Velasco expect that environmental-management standards will be achieved before air-emission standards because the former are less dependent on capital investment. However, Congress now supports the new Environmental Framework Legislation, and the authors report that 1990 was a watershed for public companies' setting realistic and effective regulatory goals.

In Chapter 4, Alfredo Núñez-Barriga, a mining engineer with training in development studies, examines the range of environmental problems of the diverse mining industry in Peru and investigates their site-specific and policy-related factors. He concludes that ownership — private, domestic, foreign, or state — is not a key explanatory factor in environmental performance, whereas the time in operation or, as he calls it, the "longevity of production capabilities," is. Centromín Perú S.A., which is more than 100 years old and has experienced periods of foreign and state ownership, illustrates this well.

Núñez-Barriga also argues that the vintage of technology is a key factor: the older the technology the greater the pollution problem is likely to be and the more radical and costly the solutions for those pollution problems are likely to be. Núñez-Barriga also reports that size of the firm fails to explain poor environmental practice, if pollution per unit of production is considered. Núñez-Barriga makes some interesting remarks regarding the relationship between mineralogical complexity and environmental pollution. He argues that there are barriers to the acquisition and transfer of clean technology that relate to the polymetallic nature of many Peruvian mineral ores.

This Peruvian case study illustrates the recent but growing environmental awareness of the country's main production enterprises across a range of minerals and regions. Most important, it highlights that change has come about through industry and state collaboration, rather than through government imposition of unrealistic regulations, with costly compliance and potential bankruptcy for some firms. Under the new legislative regime, most companies have developed site-specific environmental management and adequation programs. However, Núñez-Barriga reports that the mining industry in Peru tends to rely on established, external consultants to undertake environmental assessments and the planning of environmental-impact-mitigation measures, even where local capacity for this exists.

In Chapter 5, Maria Hanai, a sociologist, compares the economic roles of formal and *garimpo* gold mining in Brazil. She analyzes their environmental impacts and draws lessons for their mitigation. Hanai highlights the economic importance of the *garimpo* sector during the 1980s and its decline relative to formal gold mining during the 1990s, with *garimpo* mining supplying about 75% of gold production in the late 1980s and less than 50% in the early 1990s.

Hanai examines relationships among gold-mining techniques and their environmental implications. Small-scale *garimpo* mining is particularly polluting, with the hazards of mercury use and frequent large-scale land and watercourse degradation. Notwithstanding, she reports a fundamental link apparent throughout South America between poverty, or socioeconomic context, and environmental practice. This appears in the constraints these miners face acquiring credit to invest in improved technologies and in the lack of opportunities for education that also inhibits their adoption of improved environmental-management practices.

Hanai makes some policy recommendation as a basis for further research. These include technology-policy initiatives for education and innovation incentives for *garimpo* mining, as well as the adaptation of mining legislation to incorporate *garimpo* gold production.

In Chapter 6, Teresinha Andrade, a minerals-technology researcher, examines the environmental issues in Brazilian tin production. She also describes an industry divided between *garimpos* producers and mining companies and analyzes the different environmental problems with each type of production. By relating them to evolving environmental legislation, she identifies areas of potential convergence and conflict in the future relationship of tin mining to the environment. She reports that the industry is under pressure and has few resources for environmental concerns. Because of the low price of tin on the international market following its price collapse in the 1980s and because of competition from cheaper tin from China, the Brazilian tin industry has difficulty implementing the new environment-recovery plans and, for the most part, concentrates on the remediation of land degraded through past tin mining in response to regulatory pressures, rather than proactively responding to societal pressures.

Andrade comments that the most serious pollution problem is nonpoint-source pollution across mining regions, such as the silting up of large tracts of rivers. Scientists are reporting irreversible ecological degradation, the modification of gene banks and profiles of animal and plant life, changes to the soil structure, and new incidences of pests and disease. Andrade notes that current environmental legislation is retrospective and focuses on cleaning up existing pollution with current technology. She recommends emphasizing policy mechanisms to stimulate pollution prevention from the outset by providing incentives for technological and organizational change. On the basis of her interviews, she argues that this would be more attractive to mining companies, and she argues that a more democratic and regional approach to developing environmental policies is needed to find integrated and lasting solutions to the environmental impacts of Brazilian tin mining.

Liliana Acero reports, in Chapter 7, on the bauxite, alumina, and aluminum industry in Brazil. Her objectives are to document various companies' managerial approaches to the environment; to relate these practices to recent laws and regulations regarding environmental controls and planning; to measure and illustrate changes in the ways companies' environmental practices respond to specific new environmental legislation; and to describe some of the environmental effects experienced by local communities. Acero argues that regulating the environmental practices of transnational bauxite, alumina, and aluminum producers, either locally or in their home countries, is a necessary but not sufficient condition for effectively implementing sound environmental policy at the operational level. She finds that some regulations are adhered to more systematically than others and that this relates more to the economic benefits that accrue to the company than to either government or societal pressures, unless the environmental problem is very visible and has triggered a specific public response. Acero attributes some environmentally proficient practices of some transnational firms to their greater technological capacity and financial resources, although some transnational operations are not at the forefront of environmental proficiency in Brazil. She favours the strategy of Companñía Vale do Rio Doce S.A. (the state mineral producer) for reforesting and rehabilitating mined lands, which she argues is superior to the strategies of the international firms. Acero also describes lag phases in local implementation of practices already adopted in the companies' more stringently regulated home countries. She describes loopholes in local laws and regulations, along with failures in their effective implementation, which means that companies need to be proactive to achieve sound environmental track records. Acero concludes that environmental soundness depends not only on effective environmental regulation and efficient technical choices but also on the institutional context; the support, if any, given to environmental policies; and having the educational capacity and political interests needed to operationalize the law. Without the latter, Acero asserts, neither regulation nor technical or managerial solutions are sufficient to achieve truly environmentally sensitive minerals production.

Finally, in Chapter 8, Fernando Loayza analyzes, from a minerals-economics perspective, the links between competitiveness, environmental performance, and technical change in the Bolivian mining industry. He develops a dynamic economic model of the mining firm, which is empirically tested in a multiple-case study of four Bolivian mining companies and seven mining operations. This model combines an economic theory of depletion and a theory of pollution and relates investment behaviour to pollution per unit of output. It starts from the assumptions that companies compete through technical change and that competitive companies can increase their production capacity and technological capability over time.

The significance of Loayza's study is its theoretical and empirical demonstration of how mining firms' dynamic efficiency affects the internalization of environmental-damage costs. Dynamic efficiency — the ability to innovate and gain economies of scale — is not only a significant influence on a firm's ability to compete but also a principal determinant of its environmental performance. Because increased competitiveness encourages investment in technological capability and production capacity, an improvement in competitiveness tends to reduce pollution per unit of output, whereas a decline in competitiveness tends to increase pollution per unit of output. Thus, Loayza's analysis illustrates how pollution results not only from a market failure to adequately price environmental resources but also from a lack of dynamic efficiency within firms. The implication for environmental policy is that regulatory initiatives to reduce pollution should both consider mechanisms to make firms internalize externalities and address the inefficiencies of some firms, along with the dynamic capacities of others. Policy mechanisms should promote both environmental proficiency and economic efficiency.

The methodology of Loayza's study justifies applying its conclusions beyond Bolivia to other mineral-producing countries.

Conclusion

On a final note, the environmental imperative has been gaining momentum in recent years, along with the liberalization policies of Bolivia, Brazil, Chile, and Peru. Therefore, by the time these studies are published, some of the regulatory initiatives they describe may be out of date. Readers are therefore urged to view this book not only for the information it contains but also for its value as an historical document that, through empirical investigation, challenged some of the conventional wisdom that surrounded the dawn of the environmental imperative.

It should also be seen as a record of a process of research and education that in turn highlights the advantages of working together across both disciplines and national boundaries to ensure that future paths are forged to more environmentally sustainable development. We hope that MERN has contribute to ensuring a more sustainable future.

CHAPTER 1[1]

ENVIRONMENTAL REGULATION, INNOVATION, AND SUSTAINABLE DEVELOPMENT

Alyson Warhurst

More than two decades ago, the famous "Report to the Club of Rome: The Limits to Growth" (Meadows et al. 1972) predicted that the principal problem facing the world would be the depletion of nonrenewable resources, notably fossil fuels and metals. It was projected that tin, for example, would run out in 1987. However, that year saw an oversupply problem in tin markets, and several mines closed. Indeed, with technical change, recycling, and the discovery of new oil and mineral reserves, those predictions have proven to be false. The Meadows et al. report stimulated a lively debate. For example, the Science Policy Research Unit (United Kingdom) argued that institutional change and a change in the world research and development (R&D) system, and therefore in the rate and direction of technical change, could avert the predicted crisis (Cole et al. 1973; Freeman and Jahoda 1978).

In the last decade, the environmental debate has focused on the depletion and degradation of renewable resources, such as water and air. Consequently, the term *sustainable development* has been used to reflect a growing concern about the interaction between economic activity and the quality of the environment. The 1987 Brundtland Report, of the World Commission on Environment and Development, defined *sustainable development* as "development that meets the needs of the present without compromising the ability of future generations to meet their

[1] The author gratefully acknowledges Shirley Crawford and Gill Partridge for kindly assisting with the preparation of this paper and Kathleen Anderson, Rod Eggert, and Richard Isnor for providing reference materials and feedback. The empirical research was supported by a grant to the Mining and Environment Research Network from the John D. and Catherine T. MacArthur Foundation. Parts of this paper build on a more detailed study, *Environmental Degradation from Mining and Mineral Processing in Developing Countries: Corporate Responses and National Policies* (Warhurst 1994), published by the Organisation for Economic Co-operation and Development.

own needs" (WCED 1987, p. 43). This implies that economic policy should encompass environmental conservation and that the goal of more equitable economic growth refers to both intergenerational and geographic equity (Jacobs 1991).

Leaders of the G7 (group of seven leading industrialized nations) adopted the principle of sustainable development at the Toronto Summit in 1988 (Jacobs 1990, p. 59), and the 1992 Earth Summit in Rio de Janeiro heralded a more global commitment to its aims. However, the widespread adoption of the principle by policymakers, academics, industrialists, and environmentalists has not yet led to a systematic effort to make it operational through measurable policy targets or policy mechanisms for implementation. Nonetheless, regulation is slowly moving in this direction.

Previous policy, guided by the polluter-pays principle, dealt mainly with the results of environmental mismanagement — pollution — and its treatment after it occurred. The new regulatory principle — pollution prevention pays — aims to promote competitive and environmentally sustainable industrial production. This paper argues that successful implementation of the pollution-prevention principle will require the introduction of new policy mechanisms designed to both stimulate technological innovation in firms and encourage the commercialization and diffusion of those innovations across the boundaries of firms and nations. This means that government efforts to promote and regulate industry, which have traditionally been separate efforts, will need to be combined (Warhurst 1994).

This paper analyzes the challenge to policymakers posed by pollution-prevention approaches to environmental management. It develops the concept of corporate environmental trajectories for evaluating the relationship between regulatory regime and competitiveness and the implications for sustainable development. It then discusses policy mechanisms that may be used to stimulate the development and diffusion of clean technology. (The term *clean technology* is used here to refer to industrial processes that incorporate current best practice into environmental management. The term is not intended literally; indeed, a more accurate term would be *cleaner technology*.) Case studies of mining operations around the world, drawn from the research of the Mining and Environment Research Network (MERN), are used to illustrate these arguments. (The term *mining* is used here to cover all aspects of metals production, including mine development, extraction, smelting, re-mining, and waste management.) The paper shows how policy guided by the pollution-prevention principle represents an advance over previous policies guided by the polluter-pays principle. However, the paper highlights two flaws in pollution-prevention regulatory approaches: first, the firms that pollute the most are mismanaging the environment precisely because of their inability to innovate;

second, the most efficient firms are generally better environmental managers because they are innovators and are able to harness both technological and organizational change to reduce the production and environmental costs of their operations. The paper concludes by suggesting a new policy principle: environmental innovation.

This analysis recognizes that mining is a highly heterogeneous activity and that winning metals requires the removal and processing of vast quantities of rock (Winters and Marshall 1991; Tilton 1992). Some pollution can clearly be prevented, and the inevitable by-products of mining can be treated, recovered, or recycled. Radical technological and organizational innovation can change the broader context of metals production and the resulting pollution.

Although improving the environmental management of the mining industry production is the primary focus of this paper, the author recognizes that this is only one objective of sustainable development. Policy also needs to address poverty, education, health and welfare, the agricultural sector, and rural development. Nonetheless, the analysis may have implications for other industries for which institutional change, technology transfer, and training are requirements for sustainable development.

The policy challenge of pollution prevention

To meet the pollution-prevention principle requirement that pollution be reduced at source, firms must either change their technology or reorganize their production process, or both. To accomplish this, firms may need to develop new technological and managerial capabilities, form technological alliances with equipment suppliers, and collaborate with R&D institutions, which may in turn require policy mechanisms not currently part of pollution-prevention thinking.

The reasons for this are rooted in the determinants of environmental-management practices in the firm. Indeed, MERN's research suggests that the environmental performance of a mining enterprise is more closely related to its innovative capacity than to the regulatory regime under which it operates (Lagos 1992; Acero 1993; Lin 1993; Loayza 1993; Warhurst 1994). Capacity to innovate is in turn related to the entrepreneurial characteristics of the firm's management; to the firm's access to capital, technological resources, and skills; and to the broader policy and economic environment in which the firm operates. This suggests that technical change that is stimulated by the environmental imperative reduces both production and environmental costs, to the advantage of those dynamic companies with the competence and resources to innovate. Such companies include mining enterprises in developing countries as well as transnational firms. However, the evidence is strongest for large, new investment projects and greenfield sites. In older, ongoing operations, environmental performance correlates closely with production efficiency, and environmental degradation is greatest in operations working with obsolete technology, limited capital, and poor human-resource management. Developing the technological and managerial capabilities needed to bring about technical change in such organizations would clearly lead to more efficient use of energy and chemical reagents and to higher levels of metals recovery. Thus, improved production efficiency would result in improved overall environmental management, including better workplace health and safety.

International standards and stricter environmental regulations may pose no significant economic problems for new mineral projects, but major costs and challenges may be involved for older, inefficient operations. Controlling pollution problems in many of these cases requires costly add-on solutions: building water-treatment plants, strengthening and rebuilding tailings dams, investing in scrubbers and dust precipitators, etc. Furthermore, in the absence of technological and managerial capabilities, there is no guarantee that pollution-control equipment — environmental hardware — will be incorporated or operated effectively in the production process. Crandall (1983) found that a significant fraction of mandated pollution-control equipment was never even installed. In some instances, regulatory requirements are leading to shutdowns, delays, cancellations, and reduced competitiveness. When mines and facilities shut down, the cleanup costs are frequently transferred to the public sector, which, particularly in developing countries, has neither the resources nor the technical capacity to deal effectively with the resulting problems. In most countries (except perhaps the United States), the lack of retrospective regulation means that the pollutee-suffers-and-pays principle is alive and well and would continue under a pollution-prevention regime, unless, of course, the new policy fosters improved production efficiency and stimulates innovative capacity.

Environmental innovators

Although some mining companies have resisted environmental regulation of their existing operations, a growing number of dynamic, innovative companies are making new investments in environmental management. This is partly because these firms see an evolution toward stricter environmental regulation and because pushing forward the environmental and technological frontiers is to their competitive advantage. Being free of investments sunk in pollutant-producing, obsolete technology or having significant resources for R&D and technology acquisition, these firms develop cleaner process alternatives or select new or improved technologies from mining-equipment suppliers (who are themselves busy innovating). New investment projects increasingly incorporate economic and environmental efficiencies into the production process, not just through new plants or equipment but also through improved management and organizational practices. Some examples of dynamic environmental innovators are discussed below in three categories: smelter emissions, gold extraction, and waste management.

Smelter emissions

INCO LTD — At one time one of the world's highest-cost nickel producers, Inco Ltd was until recently the greatest single point source of environmental pollution in North America. This was due to its aged and inefficient reverberatory-furnace smelter, which emitted excessive quantities of SO_2. Inco had done all it could to improve the efficiency of this obsolete technology through incremental technical change when the Ontario Ministry of Environment introduced an intensive SO_2-abatement program to control acid rain. These factors prompted Inco to invest more than $3 billion in a massive R&D and technological innovation program (Aitken, personal communication, 1990[2]). Indeed, more than 12% of Inco's capital spending during the 1980s and early 1990s was for environmental concerns (Coppel 1992). Under the Canadian acid-rain-control program, Inco was required to reduce SO_2 emissions from its Sudbury smelter complex from 685 000 t/year to 265 000 t/year by 1994, a 60% reduction. To achieve this reduction, Inco planned to spend $69 million to modernize its milling and concentrating operations and $425 million for smelter-$SO_2$ abatement. The modernization process included replacement of its reverberatory furnaces with innovative oxygen-flash smelters and the construction of a new sulfuric acid recovery plant and an additional oxygen plant. By incorporating two of the flash smelters, the company reduced emissions by more than 100 000 t in 1992, and by 1994 the firm expected to achieve the government target levels. Inco is now one of the world's lowest-cost nickel producers (Warhurst and Bridge 1997). Furthermore, like other dynamic companies responding to environmental regulations through innovation, Inco seeks to recoup R&D costs by aggressively licencing its technology to firms in other copper- and nickel-processing countries.

KENNECOTT CORPORATION (UTAH) — Kennecott Corporation (RTZ Group) recently launched a new smelter project in Utah. The dual aim of the project is to set a new emissions standard for smelters worldwide and to improve cost efficiencies in the processing of its ore. Advantages include the capture of 99.9% of sulfur off-gases (previous levels were 93%). Sulfur dioxide emissions will be reduced to a new world-best-practice level of about 200 lb/h (1 lb = 0.454 kg),

---

[2] R. Aitken, Inco Ltd, personal communication, 1990.

less than 5% of the 4 600 lb/h permitted under Utah's clean-air plan. The investment of 880 million United States dollars (USD) resulted in 3 300 new construction jobs and the transfer of 480 million USD to local companies through project-development contracts. The proposed Garfield smelter will expand the copper-concentrate-processing capability to the level of mine output (about 1×10^6 t of copper concentrate per annum) at about half the previous operating cost. It represents the first application of oxygen-flash technology in the conversion of copper matte to blister. (Details are from Kennecott Corporation [1992] and Emery [personal communication, 1992[3]].) The two-step copper-smelting process starts with smelting furnaces, which separate the copper from iron and other impurities in a molten bath, followed by converting furnaces, which remove sulfur from the molten copper. A new technology, known as flash converting, will be used in the second step.

Kennecott developed this unique technology in cooperation with Outo-kumpu, a Finnish company and a leader in the supply of smelting technology. Essentially, the new technology eliminates open-air transfer of molten metals and substitutes a totally enclosed process. Flash converting has two basic effects: it allows for a larger capture of gases than the current open-air process; and it allows the smelter's primary pollution-control device — the acid plant — to operate more efficiently.

The smelter will include double-contact acid-plant technology that will improve the capture of SO_2 gases as acid. The new smelter will have other environmental benefits. An extensive recycling plan will reduce water usage by a factor of four. Pollution prevention, workplace safety and hygiene, and waste minimization will be incorporated in all aspects of the design. In addition, the smelter will generate 85% of its own electrical energy by using steam recovered from the furnace gases and emission-control equipment. This eliminates the need to burn additional fossil fuel for power. The new facility will require only 25% of the electrical power and natural gas now used per tonne of copper produced.

Gold extraction

HOMESTAKE'S MCLAUGHLIN GOLD MINE (CALIFORNIA) — Opened in 1988, Homestake's McLaughlin gold mine is a good example of a mine and processing facility designed, constructed, and operated under the world's strictest environmental regime (see Warhurst 1992c). Environmental efficiency is built into every aspect of the mining process. The McLaughlin site is notable for its innovative process-design criteria, its fail-safe tailings and waste-disposal systems, and its

---

[3]   A. Emery, technical executive, RTZ Group, personal communication, 1992.

extensive, ongoing mine-rehabilitation and environmental-monitoring systems. The mining operation, with its myriad of innovative technologies, defines best practice in environmental management.

The most interesting conclusion drawn from site visits and discussions with the firm's environmental officers is that most of these environmental-management initiatives have resulted in no substantial extra cost; indeed, many have improved the efficiency of the mine and made the overall operation more economical. An extensive environmental-impact analysis was undertaken before the mining started. All plant and animal species were identified and relocated in readiness for site rehabilitation once the mining operations end. Air, soil, and water quality were measured in detail and water-flow patterns were determined to provide the baseline for future monitoring programs. Assaying was done not just of the gold ore but also of the different types of gangue material so that waste of different chemical compositions could be mined selectively and dumped in specific combinations to reduce acid mine drainage. Local climatic conditions were evaluated to determine the frequency of water spraying needed to reduce dust; evaporation rates were evaluated to assist in water conservation and to determine the flood-risk potentials of tailings ponds. The tailings ponds were constructed on specially layered, impermeable natural and artificial filters, with high banking to prevent overflow and with secondary impermeable collecting ponds for use in the rare case of flooding.

At many other mining projects, site rehabilitation is seen as a costly task to be undertaken at the end of a mining operation, but at the McLaughlin gold mine rehabilitation began immediately and is an ongoing activity. This not only spreads expenditure more evenly over the life of the mine but also allows more efficient use of truck and earth-moving capacity and of construction personnel. When waste piles reach a certain size, soil (overburden previously stripped from the mine area and stored) is laid down and revegetation begins. Although mining at McLaughlin had been under way for only 3 years at the time of writing, extensive areas of overburden and waste had already been successfully revegetated, immediately reducing environmental degradation and negative visual impacts. In addition to these in-built environmental controls, Homestake has sophisticated environmental monitoring in place. Seepages, emission irregularities, and wildlife and vegetation effects can be detected and rectified immediately, reducing the long-term risk of expensive shutdowns, costly court cases (for water toxicity, for example), and the need for treatment technologies.

Waste management

In the minerals industry, marginal-ore dumps, tailings, and the removal of over-burden result in considerable quantities of waste rock (Gray 1993). Any toxicity associated with that waste is principally a direct result of the loss of expensive chemical reagents or of metal values. Public policy has not yet taken up the challenge of promoting R&D geared toward waste-toxicity reduction or waste-treatment innovations. One interesting area of research is the application of bio-technology to waste treatment (Warhurst 1991a).

The task of improving environmental-management practices within ongoing mining operations may not be adequately supported, however, if waste treatment is considered a third priority, below pollution prevention at source and recycling. (For an explanation of the federal approach to pollution prevention in the United States, see the chapter on pollution prevention in *Environmental Quality: The 23rd Annual Report of the Council on Environmental Quality, Together with the President's Message to Congress* [CEQ 1993].) This again suggests that pollution-prevention policy needs to focus on the environmental-innovation process, rather than strictly on pollution reduction at source.

The following two examples demonstrate how innovation can reduce pollution. The approach taken — the integration of waste management into the production process — does not always represent an add-on regulatory cost as perceived by conventional wisdom; indeed, there are competitive advantages, as well as environmental benefits, to using such a strategy.

WATER TREATMENT AT HOMESTAKE'S MINE AT LEAD (SOUTH DAKOTA) — Facing regulatory pressure to clean up a cyanide seepage problem, Homestake was able to turn the situation to its own advantage. Its R&D staff developed a proprietary biological technique for treating the effluent, which led to the recovery of local fisheries and water quality in the mine's vicinity at Lead, South Dakota (Whitlock and Crouch 1990). To recoup and profit from its investment in R&D, the company is now actively commercializing the technology, which could be widely applied at other gold-leaching plants.

WATER TREATMENT AT EXXON'S MINE AT LOS BRONCES (CHILE) — Exxon is expanding its mining project at Los Bronces, Chile, into one of the largest open-pit copper mines in the world. The expansion will result in the stripping off of very large quantities of overburden and in the creation of low-grade ore dumps. Before the mine was developed, the Chilean government warned Exxon that it would be imposing financial penalties for the water-treatment costs for the expected acid mine drainage into the Mantaro River, the source of Santiago's drinking water.

This warning became the economic justification for a bacterial-leaching project at the mine. A feasibility study showed how profitable it can be to leach copper from waste at the same time as preventing otherwise naturally occurring pollution (acid mine drainage). More than 1×10^9 t of waste and marginal ore below the 0.6% Cu cut-off grade is expected to be dumped during the project's life span. The waste could have an average grade of 0.25% Cu and would therefore contain a lucrative 2.5×10^6 t of metal, worth about 3.5 billion USD at 1985 prices (Warhurst 1990). The study demonstrated that with a 25% recovery rate, high-quality cathode copper could be produced profitably at 0.39 USD/lb by recycling mine- and dump-drainage waters through the dumps over a 20-year period.

This was shown to have the double advantage of extracting extra copper and avoiding government charges for water treatment. Both investment and operating costs were less than two-thirds of estimated costs for a conventional water-treatment plant, which would not have generated saleable copper. The Los Bronces mine demonstrates the potential economical benefits of building environmental controls into a mine at its development.

These few examples suggest that dynamic companies are not closing down, reinvesting elsewhere, or exporting pollution to developing countries with less-restrictive regulatory regimes. Rather, these companies are adapting to environmental regulatory pressures by innovating, improving, and commercializing their environmental technology and environmental-management practices, at home and abroad.

Trends in distributing environmental costs: from "pollutee suffers," to "polluter pays," to "pollution prevention pays"

Environmental regulation is frequently seen as a way to distribute the environmental costs of industrial production. Its aim is to shift the cost burden of environmental mismanagement from the pollutee to the polluter.

According to conventional wisdom, two types of costs are incurred in industrial production (Tilton 1992): the internal costs of labour, capital, and material inputs, which the company pays; and the external costs of environmental damage, such as ecological degradation, water pollution, and air contamination, which the company does not pay. This analysis runs the danger of assuming that a fixed cost is associated with each increment of pollution and that the reduction of this cost burden to society will result in a corresponding incremental increase in the firm's production costs. Tilton (1992) described this view clearly in his account of the

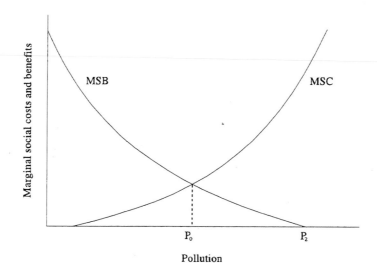

Figure 1. The marginal social costs (MSC) and marginal social benefits (MSB) of pollution.
Source: Tilton (1992).

relationship between the marginal social benefits (MSB) and marginal social costs (MSC) of industrial production, with pollution as an externality (Figure 1). The argument rests on the assumption that the socially optimal use of an environmental resource occurs when the additional benefits (in terms of goods and services it derives by permitting one more unit of pollution) equal the additional costs it incurs. In economic terms, this is the point at which MSB = MSC. If all social costs and benefits of pollution are incurred or internalized by the producing firm, the firm will have an incentive to pollute only up to this optimal point (P_o). However, if the firm realizes all the benefits associated with pollution, but not the costs, it has an incentive to expand its production until the additional benefits from causing a further unit of pollution are equal to zero. Note that in this circumstance, pollution has reached P_a, which is far beyond the optimal point, P_o.

The cost burden of this falls on society. Indirectly the pollutee pays, although the state may absorb these costs to a degree. Furthermore, as consumers do not pay the full social costs of production, pollution-intensive goods are usually underpriced and, consequently, overproduced and overconsumed. It is then argued that this situation can lead to production inefficiencies because "free" environmental resources may be substituted for labour, equipment, and other inputs (for which the firm must pay). For example, a firm may engage in the excessive and damaging use of water resources, rather than incorporating a treatment and recycling plant for effluent. This in turn reduces the entrepreneurial capacity of the firm and, most importantly, acts as a disincentive to innovate. Such a sequence of events explains in part the decline of Bolivia's state mining company, Corporación Minera de Bolivia (Comibol), and its related mismanagement of the environment (Jordan and Warhurst 1992; Loayza 1993). However, central to this idea is the assumption of a fixed environmental cost, to be either externalized or internalized. This paper challenges this assumption, arguing that technological change can reduce the environmental costs of production.

External environmental costs

For policymakers, estimating the costs of natural-resource degradation associated with mineral exploitation is a complex task. The most significant problem is devising ways to share these costs among the polluter, state, and community. Such costs are high, particularly with old and ongoing operations.

In the past, environmental costs were largely measured in terms of remedial treatment of degraded water, investment in environmental-control technologies, or compensation for damage caused to local farmland by toxic dust. More recently, environmental costs have been estimated in terms of extensive rehabilitation of the former mine and plant site for alternative uses; such rehabilitation could include revegetation or the construction of leisure facilities (Kopp and Smith 1989). However, in developing countries, it has been argued, the mining industry has traditionally been structured to externalize such environmental costs so as to maximize profit — the industry appropriates undervalued resources and shifts the environmental costs to others, rather than improving efficiency and innovating.

When it comes to evaluating these costs, it should be remembered that those most affected by environmental pollution from mining in developing countries are generally those least able to understand and respond to it — remote miners' families or isolated rural communities. Responses are typically short term and nonsustainable. For example, when farmlands were ruined by pollution from the Karachipampa tin-volatilization plant in Bolivia, the peasants were offered small compensation payments covering only the loss of particular harvests, rather than the potential loss of their livelihoods. In contrast, in the United States, the fastest growing area of consultancy is in the assessment of liability for natural-resource damage — propelled by the *Comprehensive Environmental Response, Compensation, and Liability Act* and *Superfund Amendment and Reauthorization Act*, or "Superfund" laws, which apportion blame for environmental damage to any one of a mine's past owners and charge them with the cost of government contracts to clean up and rehabilitate the damaged site.

Inevitably, some environmental degradation results from mining. Although such pollution has a negative economic impact, it often presents unrealized economic opportunities — for firms, as well as for society. For example, toxic by-products that could be economically reprocessed are frequently dumped instead. This is the case especially in developing countries, where inaccurate sampling or inefficient technologies result in such loss. Mining high-grade ore and dumping low-grade ore may be a short-term expediency for boosting foreign-exchange earnings in times of crisis, but it results in greater environmental degradation (higher risk of acid mine drainage from dumps) and the loss of long-term revenue. Water-treatment projects are often instigated at the time of mine closure, which is more costly than preventing acid mine drainage from the outset. Such pollution control could result in the recovery of metal values, through solvent-extraction or electrowinning techniques. Finally, some companies have had to pay the health-care costs for communities that drink degraded water, which are in many cases greater that the cost of the technical change needed to treat the chemical effluent in the first place. Even if firms are forced to absorb some of the environmental costs of their operations in the long term, this does not necessarily translate into improved efficiency in the short term.

Considerable work is still needed to quantify the nature and extent of environmental degradation caused by metals production. Currently, only isolated case studies exist, and little systematic analysis of the problem has been undertaken. It is difficult to generalize because local geological, geographic, and climatic conditions affect mineral and ore chemistry, soil vulnerability, and drainage patterns and hence the extent of the environmental hazard. Furthermore, the degree of environmental hazard is affected by the social and economic organization of the production unit, including such factors as the firm's size, history, and ownership structure, as well as its propensity to innovate.

The polluter-pays principle and the internalization of environmental costs

A combination of political, economic, and environmental elements has given rise to the polluter-pays principle, which in essence requires polluting companies to internalize the external costs of environmental damage caused by their production of goods and services. Member countries of the Organisation for Economic Co-operation and Development have endorsed this principle for many years, and the 1992 deliberations of the United Nations Conference on Environment and Development (UNCED) heralded commitment to its application on a more global scale.

The norm for environmental regulations incorporating the polluter-pays principle is for governments to set maximum permissible discharge levels or minimum levels of acceptable environmental quality. Such command-and-control mechanisms include Best Available Technology (BAT) standards (including Best Available Technology Not Entailing Excessive Costs standards), clean-water and clean-air acts, Superfund laws for determining cleanup and liability, and a range of site-specific permitting procedures, which tend to be the responsibility of local government within nationally approved regulatory regimes. Implementation and enforcement of command-and-control mechanisms tend to be the responsibility of administrative agencies and judicial systems. However, such polluter-pays regulations may not promote real reductions in environmental degradation or improve environmental management in metals production on a broad scale.

First, the polluter pays only if discovered and prosecuted. This requires technical skills and a sophisticated judicial system, often activated only after the pollution problem has become apparent and caused potentially irreversible damage. Moreover, in developing countries, serious economic and political constraints limit the implementation of environmental regulations and the penalization of polluters (Warhurst 1994). For these reasons, environmental regulations tend to address the symptoms of environmental mismanagement (pollution problems), rather than the causes (economic and technical constraints; lack of access to technology or information about better environmental-management practices). This neglect can be serious because for certain types of pollution, such as acid mine drainage, it is extremely costly and sometimes technically impossible to trace the cause and thus to rectify the problem and prevent its recurrence. Certain environmental controls may only work if incorporated into a project from the outset (such as buffer zones to protect against leaks under multitonnage leach pads and tailings ponds) or if combined with economic incentives.

Second, a plant may meet BAT standards at start-up without being able to achieve the specified effluent and emission levels throughout its life span. Technical problems may arise; cumulative production inefficiencies are not unusual; and the quality of concentrate or smelter feed may change if supply sources are changed. Moreover, the site-specific nature of mining operations has serious implications for monitoring, as technology has to be fine-tuned for each mineral deposit. It would also be erroneous for a regulatory authority to assume that standards are met simply because a preselected item of technology has been installed. Ongoing management and environmental practices at the plant are as important as technical hardware in achieving environmental best practice. Evidence from MERN research suggests that these problems are endemic to metals production in many developing countries (Núñez 1992; Hanai 1993; Loayza 1993; Warhurst 1994), where obsolete technology is widely used without modern environmental controls or safeguards. In the industrialized countries, new concentrators and roasting plants tend to be computerized. Automatic ore assaying and in-stream analysis give an accurate picture of the chemical composition of the ore feed, information that is needed for fine-tuning the pressure, heat, cooling, and environmental-control systems and for accurately predicting and monitoring emissions. However, where these controls are missing and, in particular, where ore feeds are of variable composition (in terms of, for example, sulfur, lead, and arsenic content), the pollutant effects of emissions also vary. Furthermore, pollution increases with the inefficient or excessive use of fossil fuels, particularly by poorly lagged roasters, inefficiently operated flotation units, and energy-intensive smelters. It could, therefore, be argued that command-and-control regulatory instruments are unlikely to result in a reduction of pollution, as they do not alter the capacity of a debt-ridden and obsolete mining enterprise (especially in developing countries) to implement technical change. Such a firm might find it preferable to risk detection, pay a fine, or mask its emission levels than to face bankruptcy from investing in new technology while its capital is scarce.

Third, BAT standards and environmental regulations of the polluter-pays type tend to presume that technology is static — they're based on a technology that was best at one time. Such regulations act as a disincentive for equipment suppliers, mining companies, and metal producers to innovate. Or perhaps they have innovated, but their innovations, which may have required substantial R&D resources, have been superseded by a regulatory authority's decision about what constitutes BAT for their activity. Ashford and Heaton (1979) described instances where the use of environmental innovations that were superior to the specified BAT was discouraged because the regulators were unfamiliar with their design or operation.

Regulations obliging the polluter to pay tend to lead to end-of-pipe, add-on, or capital-intensive solutions (such as smelter scrubbers, water-treatment plants, and dust precipitators) for existing technology and work practices, rather than promoting alternative environmental-management systems and technological innovation. Moreover, if regulations are incremental, they may promote technical change that is also incremental, involving the addition of numerous new controls at greater cost and with more overall degradation than if a new, more radical change had been introduced in the first place (see Kemp and Soete 1990; Freeman 1992). Such regulations may also require specific reductions in pollution without regard to cost or local context. The regulations may refer to the chemical composition of an effluent in isolation, disregarding the site-specific precipitation, evaporation, or soil and geological conditions that affect the discharge rate and pattern.

Such regulations also result in a single-medium approach; consequently, firms may respond by shifting pollution from one medium to another (such as from emissions to effluent). An interesting example of this occurred at the Alcan

Ltd bauxite mine and alumina plant in Jamaica. Foreseeing impending envi-ronmental regulations and responding to public concerns in its home country (Canada), Alcan gave support to a local university to develop an innovative solu-tion for the disposal of red-mud sludge from the bauxite mining operations. Previously, the sludge had been dumped in a large catchment pond, but toxic seepages into surrounding soils and groundwater were reported. The university developed a process called red-mud stacking, which involves sun-drying to remove much of the moisture from the sludge and stacking of the material in much less obtrusive piles. However, this technology does not address the toxic seepage of the previously dumped slurries. Nor does it offer a solution to pollution per se because it replaces water pollution to a large extent with dust pollution, which is less stringently regulated. Moreover, a change in the production process to facili-tate the recovery of caustic soda from the "mined" dry-mud stacks means that more of that chemical is discharged than with the previous method. The dust pol-lution, plus overflows from those parts of the dry-mud stacks that become water-logged during tropical rain showers, presents a greater toxic hazard than the previous low-level seepages.

Also, industry may cooperate less with this regulatory approach because the rules are continually changing and the costs of complying are increasing. Finally, such regulations ignore the human-resource contribution to sound envi-ronmental management because they emphasize a specific pollution-control tech-nology (environmental hardware), rather than training, managerial approaches, and information diffusion (environmental software).

TOWARD POLLUTION PREVENTION — Interest has been growing in the use of market-based mechanisms whereby polluters are charged for destructive use based on estimates of the damage caused. An important justification for market-based incentives is that they give companies greater freedom to choose how best to attain a given environmental standard (OECD 1991). Market-based incentives, by remedying market failures or creating new markets, may permit more economi-cally efficient solutions to environmental problems than government regulations substituting for imperfectly functioning markets would. Two categories of incentives exist (O'Connor 1991; Warhurst and MacDonnell 1992). One group, based on prices, includes a variety of pollution taxes, emission charges, product charges, and deposit–refund systems. For example, a mercury tax has been dis-cussed in Brazil; a cyanide tax, in the United States. The other group of incen-tives, based on quantity, includes tradable pollution rights or marketable pollution permits. A related measure is the posting of bonds up front for the rehabilitation of mines upon closure. This is now standard practice in Canada and Malaysia.

Few governments have designed systematic incentives for industry to innovate and develop new environmental technology. An approach such as this might change the very essence of environmental costs by no longer assuming they are fixed. Indeed, in two further areas, policy approaches may contribute to improved environmental-management practices. First, private, bilateral, and multilateral credit is frequently contingent upon the use of environmental-impact assessments and best-practice environmental-control technologies in new minerals projects. Requiring mandatory pollution-prevention plans as a condition for obtaining mine permits would be a complementary policy mechanism. A growing number of donor agencies — in Canada, Finland, Germany, and Japan, for example — are also emphasizing training in environmental management. Second, some governments are promoting R&D (jointly and within industry and academic institutions) to evaluate toxicity from mining pollution and develop cleanup solutions. For example, the Canadian government has funded R&D programs on abatement of acid mine drainage and SO_2 emissions. However, considerable scope remains for expanding these approaches, as argued below.

Pollution prevention and the demise of the environmental trade-off

The pollution-prevention principle differs from the polluter-pays principle because intrinsic to the notion of reducing pollution at source is a nonstatic vision of the environmental costs of production. The polluter-pays principle implies that firms internalize a fixed environmental cost. However, firms' pollution-prevention efforts demonstrate a diminishing intrinsic value of that environmental cost (not its shifting to others), and this leads to the demise of the environmental trade-off.

The US Environmental Protection Agency (EPA) is still defining pollution prevention in terms of internalized fixed environmental costs: "pollution prevention requires a cultural change — one which encourages more anticipation and internalizing of real environmental costs by those who may generate pollution" (Habicht 1992). Clarifying the concept of pollution prevention is important because it will inform the design and implementation of policy to achieve it.

The concept of corporate environmental trajectories (see Figure 2) illustrates the fundamentally different nature of technical change and therefore of the environmental costs of applying pollution prevention to metal-mining operations. Such trajectories describe the evolution of a firm's competitiveness in response to both changing market conditions and regulatory requirements. Governments and, indeed, corporate strategists need such policy tools to predict the environmental practices and competitive behaviour of firms under various market conditions and regulatory regimes and to identify the warning signs of declining competitiveness, impending mine closures, and their environmental effects. For example, combined regulatory and market pressures may prompt mine closure in advance of expected ore depletion. But in many countries a bankrupt firm is no longer responsible for its cleanup problem, so the burden frequently falls on the state, which has neither the resources nor the skills to deal with such a large-scale and complex problem. (See Warhurst and MacDonnell [1992] for a discussion of the case of Carnon Consolidated Ltd in the United Kingdom and numerous articles about the Summitville Mine Superfund site in Colorado.)

Technical change and corporate environmental trajectories

Characteristically, enterprises respond slowly to environmental pressures, and their responses predominantly reflect the regulatory regimes and public climate of their home countries. Their responses also depend on the nature of their operations in terms of

- The minerals;

- The level of integration of mining and processing;

- The stage in the investment and operations cycle; and

- The economic and technological dynamism of the firm (whether it has the financial, technical, and managerial capabilities to be an innovator).

After a period of using static technology, the mining and mineral-processing industry is currently going through a phase of technical change, with dynamic firms developing new smelting and leaching technologies in response to economic as well as environmental constraints. This trend is stimulated by rapidly evolving environmental-regulatory frameworks in the industrialized countries and the prospects of their application, reinforced by credit conditionality, in the developing countries. Changes in technological and environmental behaviour in this context are evident, particularly in the large North American and Australian mining firms, and are increasingly apparent in firms based in developing countries, such as Brazil, Chile, and Ghana. However, it seems that only new operators and dynamic private firms are changing their environmental behaviour; state-owned enterprises and small-scale mining groups in developing countries continue, with some exceptions, to face constraints on their capacity to change environmentally damaging practices.

Inevitably, only dynamic firms with new project-development plans are in a position to invest in R&D to develop more environmentally sound alternatives or to raise the capital to acquire them from technology suppliers. After a long period of conservative and incremental technical change, firms are developing alternative processes for mineral production that are more economical and less environmentally hazardous. Furthermore, these firms are beginning to sell their technologies, preferring to commercialize their innovations to recoup their R&D costs than to sell obsolete technology and risk shareholders' displeasure or retrospective penalties as developing countries start to enforce environmental regulations. Some mining firms have even pushed technology beyond what is required to meet existing regulations. These firms are seeking ways to increase regulation, particularly on a worldwide scale, because they can meet stricter standards and use their new environmentally sound technologies to competitive advantage.

The assumption that environmental regulation represents a cost burden to the firm is challenged by evidence that improved environmental management in mining operations need not be detrimental to economic performance and may in some cases even have economic benefits. This is intrinsic to the pollution-prevention-pays principle. Environmental regulations are here to stay, and they're bound to become more widely adopted, more stringent, and better enforced. It follows, then, that a greater share of the metals market will be lost by those firms that avoid environmental controls (only later to be forced to internalize the high costs of having done so). This share will be won by firms that get ahead of the game, play a role in changing the industry's production parameters, and use their innovative capabilities to improve their competitive advantage. Implementing lean-production practices is one example of this.

Much can be learned from the manufacturing sector about the development and success of lean production and related Japanese work methods, such as just-in-time inventory control, waste reduction throughout the system, total quality management, and statistical process control. Lean production is defined by a simple principle: eliminate all costs that do not add competitive value to a product. Secondary principles are reduce waste, minimize space, eliminate inventories, and integrate quality control into the production process. The implementation of lean production characteristically results in the reduction of managerial roles, with increased responsibility being given to engineers and workers and a concomitant increase in multitask activities (Womack et al. 1990). A study of more than 90 plants in 14 countries, representing half of the world's automobile-assembly capacity (Womack et al. 1990; see also Graves 1991), showed that lean production, used principally by Japanese companies, has significantly improved productivity, quality, product development, and model range. The average European and North

American plants required 118 and 49%, respectively, more effort than the average Japanese plant to undertake the same manufacturing activities. Such efficiencies in Japanese plants have translated into both cheaper and better-value products, leading to the rapid growth and supremacy in Western markets of Japanese firms like Toyota, Nissan, Honda, and Mitsubishi.

The implications of applying lean-production principles to mining, or of radical process innovations with similar effects, can be expected to be remarkable. A combination of markedly lower investment and production costs and the halving of mine-development times and mine life spans would significantly affect the competitive structure of the industry and reduce negative environmental and social effects. Few mining companies have taken these ideas on board. Those that have considered alternative organizational methods include CRA Ltd (Australia), Homestake's McLaughlin gold mine (California), and Scuddles mine, of the Poseidon Group in Australia. Scuddles has implemented an innovative, multi-skilled approach to human-resource development at its underground mine in Western Australia (Mining Magazine 1991). Also important are management training and new work methods for engineers and miners. There is a clear relation-ship between good housekeeping at the plant site and environmental practices.

Figure 2 categorizes the environmental trajectories that different mining firms might take in response to environmental and market conditions. Companies and governments could use this diagram as a planning tool to evaluate the environmental and economic implications of different policies.

The average mining firm is competitive (that is, to the left of the threshold of economic competitiveness, X, in Figure 2). To a greater or lesser extent these firms produce environmental pollution, and to a greater or lesser extent they have internalized the cost of the environmental degradation associated with their metals production, in response to the regulatory regime they are working within. (The threshold of environmental competitiveness for a given regulatory context is also X, and company operations in compliance have environmental trajectories in the quadrants below the horizontal axis.) However, as a result of market pressures — mainly a real decline in metal prices — and their own economic inefficiencies, some of these firms are going bankrupt (on a trajectory toward quadrant B). They will leave a legacy of environmental pollution, and as happened with Comibol (in Bolivia) and Carnon (in the United Kingdom), the burden of cleanup will fall on the state and society. Other firms will respond by innovating (moving into quad-rant D), building improved economic and environmental efficiencies into the new

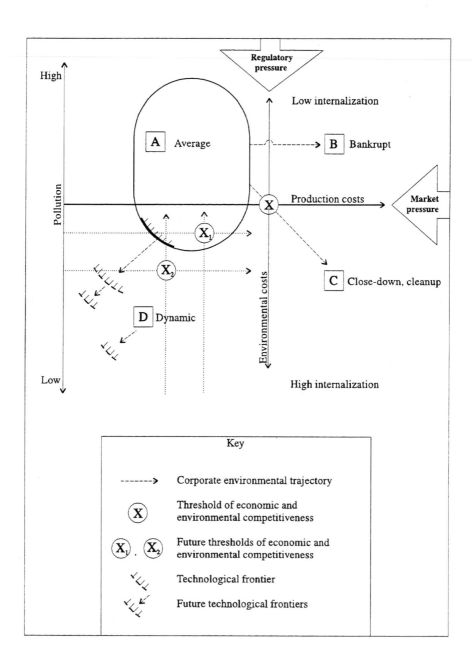

Figure 2. Corporate environmental trajectories.

generation of technology. At the same time, these innovators are protecting themselves from having to undertake more costly add-on, incremental technical change
and rehabilitation at later stages of their operation. Indeed, freed from the incum-
bent costs of retrofitting sunken investments, dynamic greenfield plants in particular can use the latest best-practice technology incorporating improved economic
and environmental efficiencies.

Nonetheless, if obliged to add on environmental controls in line with new
regulations, a growing group of firms would have to close down because the cost
of the controls and cleanup would render their operations uneconomical. The environmental trajectory of this group is toward quadrant C. Currently, examples of
this are scarce, and it is difficult to differentiate between purely environmental
factors and other factors that may be causing a firm's cost curve to increase.
However, as Figure 2 shows, that group can be expected to grow because combined market and regulatory pressures will lower the threshold of economic and
environmental competitiveness to the extent that the average firm will survive in
the new regime only if it innovates. Therefore, even previously dynamic firms will
need to keep their environmental trajectories moving ahead of the encroaching
threshold of economic and environmental competitiveness (X_1 and X_2).

These trajectories imply a serious constraint on the regulatory process for
two reasons that distinguish mining firms from their manufacturing counterparts.
First, if a mine closes down as a result of regulatory burden, its environmental
degradation may continue. Pollution in metals production is not all end-of-pipe
pollution, which stops when production ends. Rather, the closure of a mine heralds
a new phase of environmental management — decommissioning, cleanup, and rehabilitation, all of which pose significant costs. Second, in most countries, the
former operators of closed mines are not liable for the cleanup (the United States,
with its Superfund liability laws, is an exception). Therefore, pushing forward the
technological frontier and moving the threshold of economic and environmental
competitiveness may result in an overall increase in environmental degradation
(particularly where there is no liability).

The policy challenge that pollution-prevention advocates face is one of
keeping firms sufficiently dynamic to reduce their pollution at source, to profitably
clean up pollution that escapes, and in the meantime to generate increasing economic wealth. The policy challenge is therefore to promote environmental innovation. This means combining the regulation and promotion of industrial activity in
an integrated policy.

Policy mechanisms to promote environmental innovation and stimulate diffusion of these innovations

A number of important policy implications follow if one accepts the argument that the environmental practices of firms correlate most closely with their innovative capacity and that regulations are only really effective if firms have the innovative capacity to respond and change their production processes and products, including waste products. Production inefficiency and environmental mismanagement go hand in hand.

Low rates of metals recovery, high-intensity use of energy, excessive use of reagents, and so on are symptomatic of production inefficiency and are also associated with pollution, such as metal particulates, SO_2 emissions, and toxic chemical effluent. The question of economies of scale in metals production further constrains the choice of technology for minimizing waste and maximizing metals recovery.

A major implication for pollution-prevention policy is that in tackling the most significant polluters, it must also target the most inefficient (this means targeting the least innovative). Such a strategy poses a potential problem for pollution prevention: promoting innovation is key to reducing pollution at source, but the firms generating the most pollution have insufficient technological and managerial capabilities or capital resources to innovate.

A corollary of this is that the most successful metals producers use reagents and energy efficiently and have high rates of metals recovery. They are constantly engineering incremental improvements to optimize these levels. Therefore, concentrating environmental policy on pollution prevention at source might fail to optimize the potential of dynamic companies to develop and diffuse innovations to reduce pollution at any point in the life cycle of the mine and its products.

Another problem stems from the huge volumes of rock involved in mineral extraction — although the percentage releases of toxins may be small, the scale of pollution can be great because of the sheer bulk of throughput (Gray 1993). Therefore, policy needs to have a dual focus: innovation to prevent pollution at source; and innovation to promote profitable waste treatment, reagent and metal recycling, and re-mining. The latter focus may require a different range of incentives (along with the removal of disincentives to re-mine and treat waste under regulatory regimes), such as the *Resource Conservation and Recovery Act* of the United States. Environmental innovation is the key to progress on both fronts. It recognizes the need to integrate environmental regulations and promotion of industrial activity.

Consequently, two policy options are available for achieving pollution prevention. First, punitive regulation can be used to put the inefficient and most serious polluters out of business. Second, a range of policy mechanisms and incentive schemes can be developed to promote production efficiency and innovation in environmental management, focusing on the entire metals-production process, from mine development to waste management.

Although the first option is superficially attractive, two problems arise from pursuing sustainable development in this manner. One is that for many economies (particularly developing ones), minerals production is a crucial source of foreign exchange, government revenue, and direct and indirect employment. Punitive regulations thus put development objectives at risk, which in turn threatens the economic part of the sustainable-development equation. The other problem is that this strategy threatens the environmental part of that equation by pressuring firms to move into the close-down, cleanup quadrant of the corporate environmental trajectory (C in Figure 2). As noted earlier, this is an unattractive option because the pollution problems associated with decommissioning and rehabilitating mining sites can be horrendous and often become a responsibility of the state. Several studies have shown that retrospective legislation, such as the Superfund in the United States, is an inefficient mechanism for achieving the optimal use of resources for environmental protection (Portney 1991; Acton et al. 1992; Probst and Portney 1992; Tilton 1992).

The policy challenge for pollution-prevention advocates (if their ultimate aim is sustainable development) is thus to keep the mining industry dynamic enough to reduce pollution at source, profitably treat waste, clean up pollution on closure, and generate economic wealth (using best practices in environmental management) throughout a mine's life span. This paper argues that environmental legislation to support pollution-prevention goals must be underpinned by two further sets of policy mechanisms: mechanisms to promote environmental innovation; and mechanisms to stimulate the diffusion of these innovations among firms.

Although governments may only be explicitly concerned with the diffusion of innovation among firms within national boundaries, the commercialization of these innovations abroad can bring in revenue and improve the overall competitiveness of national mining firms and sectors. International organizations — including banks, donor agencies, and institutions concerned with implementing UNCED objectives — also need to consider policies to promote the international diffusion of clean technology.

This is not an argument against regulation but a recommendation for a more sophisticated approach to public policy. Such a policy would define the regulatory goals relating to both the production process and the output stream — that is, set something to aim at — and be underpinned by an informed technology policy to guide and stimulate industry along the fastest, most efficient route to those goals.

Mechanisms to promote environmental innovation

Policy mechanisms to promote environmental innovation in industry are of two types: expenditure programs to support clean-technology R&D, training in environmental engineering, and so on; and incentives to reward firms for environmental innovation.

Expenditure programs

Technology-policy mechanisms that support clean-technology development include expenditure programs funding R&D in selected areas of pollution prevention. Examples are Canada's R&D programs in acid mine drainage and biotechnology to clean up effluent. Another example is cofunded R&D projects involving industry–industry, industry–university, or industry–research-institution collaboration. Such programs could be supported through easily accessible, centrally compiled information-dissemination programs concerned with moving technological and regulatory frontiers.

A crucial factor in targeting R&D support is the innovation process in industry. Too often, policy documents conceptualize innovation as something that builds on government or university R&D and then is magically applied throughout an industry's operations. Such thinking is reflected to a certain degree by EPA's aims in the area of technological innovation. Evidence suggests, however, that in most cases innovation is industry driven, with firms drawing on research institutions and other firms for the additional knowledge, expertise, and technology needed to complement their own in-house R&D and engineering efforts (Rothwell 1992; Warhurst 1994). An important function of technology policy to promote source-reduction innovation should be to inhibit a possible tendency of firms to divert resources from conventional business R&D to compliance-related R&D. Focusing R&D on process innovation and making pollution prevention at source part of the overall effort to improve efficiency can be complementary aims.

Promoting innovation in pollution-prevention technology requires important changes in thinking. A multimedia approach is required because pollution prevention requires changes in process technology, not the addition of off-the-shelf, end-of-pipe controls that tend to shift pollution from one medium to another. This implies the need for a range of engineering skills to reduce or eliminate the pollutants at source (independently of where they may ultimately be discarded). New technology must be designed to deal with water and air quality and waste, as well as workers' health and consumer-product safety. Thus, training for R&D engineers in industry should be another target of policy mechanisms.

EPA policy documents on pollution prevention fail to emphasize training. Technology hardware is only one part of the equation. Of equal importance is organizational change. Mechanisms that foster lean and clean production are also needed.

Incentives

Taxation policy may need to be changed to promote environmental innovation. According to Ashford (1991), the United States gives taxation incentives in the form of accelerated depreciation for pollution-control equipment, thus supporting end-of-pipe pollution control. However, no similar incentives apply to investments in new production technology, with the result that dollar for dollar a firm is better off buying from an environmental-technology vendor than developing its own process changes. Direct taxation incentives can relate to investment in pollution-prevention technological or organizational change; R&D; engineering projects and training in specific areas of environmental management; and bonds posted up front for future pollution prevention or for reclamation on closure. The impact that punitive taxation on reagent or energy use will have on firms' competitiveness and behaviour requires careful consideration — the geological and chemical characteristics of each deposit are unique, and this affects energy- and reagent-consumption patterns. Operators may perceive such taxation as prejudicial and unfair.

Flexible taxation provisions that allow and even encourage industry to be innovative are needed to complement strict standards and regulatory goals. Regulators must possess an intimate knowledge of the types of gains firms make from technological change. With this knowledge, they will be able to determine how best to promote technological innovation and can adapt or ratchet regulations accordingly (Milliman and Prince 1989). Innovative firms should be able to use environmental regulations to their competitive advantage. The innovator would benefit from stricter technology-forcing regulations that stimulate other companies either to invest in new technology or to licence (or purchase) the innovator's technology (thus enabling the innovator to recoup some of its initial investment in R&D). Regulatory authorities need to be seen to respond in this way. For the informed regulator, the rate of technological advance in pollution control is probably the most useful measure of the effectiveness of environmental policies, a view held by a growing number of researchers, including Orr (1976), Kneese and Schultze (1978), and Milliman and Prince (1989). Training for regulators, including industrial experience as environmental engineers and corporate strategists, is thus an important part of the pollution-prevention approach.

Pushing the technological frontier forward will pull the thresholds of economic and environmental competitiveness deeper into quadrant D, as shown in Figure 2. Consequently, market conditions governing metals production will also change, to the innovator's advantage.

An important corollary of an incentives-to-innovate policy is that regulators give rewards for innovation. Usually the reward is a prize for sound environmental management, such as EPA's recent Environmental Leadership program to reward US innovators. However, regulators should make the reward side of the equation more sophisticated by analyzing the ways firms realize and expand commercial gains from technological innovation and technology diffusion.

Milliman and Prince (1989) analyzed five regulatory approaches: direct controls, emission subsidies, emission taxes, free marketable permits, and auctioned marketable permits. They found that direct controls, which are the most common regulatory tool, provide the least incentive for technological innovation; free permits and emission subsidies also provide little incentive. Emission taxes and auctioned permits are the best incentives because they reward the innovator through gains the firm makes from diffusing its technology to other firms, over and above the benefits the firm derives from its own application of the technology (Milliman and Prince 1989). This is not surprising, as polluters facing high costs for abatement will find it cheaper to buy permits than to reduce their emissions, and polluters with low abatement costs will sell their permits accordingly. Firms therefore have a constant incentive to cut emissions, as this allows them to sell permits. Tradable permits, unlike pollution charges, can guarantee the achievement of particular pollution targets because the authorities control the number of available permits.

Finally, on this issue, incentives are needed to stimulate auxiliary firms to develop and commercialize innovative clean-up technologies, including re-mining techniques. In developing countries particularly, the market for such technologies is vast, and donor agencies and development-assistance grants could play a key role in stimulating such investment. For example, more than two-thirds of the current mineral reserves of Bolivia are in dumps and tailings (Warhurst and MacDonnell 1992). Furthermore, in many developing countries, such as Peru, dynamic small- and medium-scale firms supplying a range of inputs to the mineral sector could, with incentives, expand their activities to the environmental arena (Núñez 1993). In the United States, liability regulations should be reassessed because the current barriers to re-mining and treating mining waste need to be removed.

Mechanisms to stimulate the diffusion of environmental innovations

Technological and managerial capabilities are required for innovating and for dealing with new and emerging technologies, and they are also vital for resolving pervasive inefficiencies if a firm's environmental-management strategy is to use existing technology. Technology transfer and technology partnership through joint ventures or strategic alliances are ways to build up technological and managerial capabilities. This is particularly pertinent in the developing countries, although such strategic alliances are emerging in all the major mineral-producing countries. Recent collaborative partnerships in environmental innovation include Outokumpu and Kennecott Corporation; Outokumpu and the Chilean state copper corporation, Corporación Nacional del Cobre S.A.; Cyprus Mines and Mitsubishi; Comalco, Marubeni Corporation (Japan), and the Chilean power company, Empresa Nacional de Electricidad S.A.; Battle Mountain (United States) and Inti Raymi (Bolivia); and Compañía Minera del Sur S.A. (Bolivia) and Compañía de Minas Buenaventura (Peru).

However, the concept of technology transfer should be broadened to include a real transfer of environmental-management capability. Technology transfer has traditionally meant a transfer of capital goods, engineering services, and equipment designs — the physical items of the investment — accompanied by appropriate training for operating the plant or equipment. Consequently, the innovative capacity of recipients is undeveloped and they remain purchasers and operators of imported plants and equipment. This is especially the case in developing countries, where many recipients become dependent on their suppliers to make changes or improvements to successive vintages of technology. Contractual conditions may reinforce this situation.

New forms of technology transfer in environmental management are needed to embrace

- The knowledge, expertise, and experience required to manage technical change — of both an incremental and a radical nature; and

- The development of human resources for implementing organizational change to improve overall production and energy efficiency and environmental management throughout plant and facility — from mine development, through production, to waste treatment and disposal.

This new concept of technology transfer emphasizes training and skills acquisition in environmental R&D, engineering, management, troubleshooting, repair and maintenance, environmental auditing, and so on.

In global industries like mining, international firms supply significant amounts of managerial and engineering expertise through joint ventures and other collaborative arrangements. These contributions are usually limited to the immediate requirements of the specific investment project or of the equipment purchased. Flows of technology may even be structured to match regulatory requirements. Cumulative command-and-control regulations tend to lead to incremental, add-on, end-of-pipe, capital-intensive technical change and therefore successive rounds of technology imports (Warhurst and MacDonnell 1992). However, empirical research in other sectors demonstrates that these contributions can be considerably increased without adversely affecting the supplier's strategic control of its proprietary technology (Bell 1990; Warhurst 1991a, b; Auty and Warhurst 1993).

Such an approach was at the heart of the strategy of China's National Offshore Oil Corporation, which required major oil companies, under technology-transfer agreements in their investment contracts, to transfer the skills needed to master selected areas of technology (Warhurst 1991b). Another interesting example is the Zimbabwe Technical Management Training Trust. It was founded by RTZ in 1982 to train South African Development Community professionals in technical management and leadership. Participants receive a combination of academic and on-the-job training in the operations both at home and overseas. Accelerated managerial training is possible through exposure to on-the-job problem-solving situations, with experienced colleagues, in a range of challenging technical scenarios.

Similar in-depth training programs, concentrating on environmental management, could be built into many of the proposed mineral-investment projects throughout the world. Preference could be given to investors and technology suppliers with proven environmental-management competence and a willingness to transfer their skills and knowledge. It cannot be overemphasized that all technology transfer and training efforts represent a cost to the supplier, and this cost must be covered to ensure optimal results. The danger of failing to budget for this cost is ending up with a training program in operational skills instead of one in technological mastery. Corporate partners, the government, and, in the case of developing countries, donor agencies or development banks can assist in financing these schemes. Moreover, governments, organizations, or firms will have more power to negotiate the objectives and scope of the programs if they have helped finance them.

Mine operators can purchase capital goods, engineering services, and design specifications through a range of well-established commercial channels; however, the market for knowledge and expertise, including training programs, is less mature. Active development of this market will reward innovators in pollution-prevention technology. Bilateral and multilateral agencies, development banks, and government organizations can play a major role in improving this. Agenda 21, one of the main outputs of UNCED, proposes two programs of relevance (Skea 1993) that can be expected to lead to greater industry involvement. One of these programs encourages interfirm cooperation, with government support, to transfer technologies that generate less waste and increase recycling. The other program, on responsible entrepreneurship, encourages self-regulation, environmental R&D, worldwide corporate standards, and partnership schemes to improve access to clean technology. Moreover, Agenda 21 (chapter 34) recognizes that for technology transfer to be effective, a substantial increase in the technological capabilities of recipient countries is required (Barnett 1993). The capacity to effect technical change, not just the skill to operate an item of environmental-control technology, will ultimately determine whether a recipient firm can build up the competence it needs to be successful in environmental management and environmental innovation. Broadening the concept of technology transfer to encompass these issues would also enable government and industry policymakers to more accurately assess barriers to the diffusion of clean technology.

Implications for policymakers

This analysis suggests the need for a two-tier policy approach. Policy concerning ongoing projects must cover the challenges of production inefficiency, its environmental consequences, and the clean-up requirements for mine closures and plant decommissioning. Policy concerning new investment and expansion projects should stipulate that environmental management and the flexibility to engage in further environmental innovation be built into the project at the outset. This requires negotiation among operators, equipment suppliers, and credit sources at the earliest stage.

The analysis of technology transfer for a pollution-prevention policy has two implications. First, if the policy is to work as a means of achieving sustainable industrial development (even if only a means for firms and countries to comply with UNCED recommendations), it must be underpinned by a technology policy with financial incentives that promote the commercialization of pollution-prevention innovations in overseas markets, including Eastern Europe and developing countries. Although this paper has highlighted some of the barriers to the diffusion of pollution-prevention technology, these barriers are not so much due to the absence or inadequacy of regulations as to a lack of technological and managerial capabilities, insufficient investment resources, information constraints, etc. It is argued here that industry, governments, and international organizations, including development banks, have an interest in overcoming these barriers. One route is fostering the real transfer of technology, as described above.

Second, environmental policy needs to be integrated with other government policies, such as those covering industry, trade, and technical assistance. Firms can learn a great deal from one another. Indeed, this is reflected in the formation of the International Council on Metals and the Environment (ICME) and the recent set of strategic alliances between leading technology suppliers and mining companies (see above). A major purpose of ICME is to promote sound environmental and health policies and practices to ensure the safe production, use, recycling, and disposal of metals. An important rationale for its establishment was the view that industry can benefit from pooling its expertise, exchanging information, promoting sound environmental and health policies, and working cooperatively and proactively with regulators, labour, and scientific and environmental groups. Great scope remains for the further diffusion of knowledge (environmental software) and technology (environmental hardware) among firms, between firms and regulators, and across firms and national boundaries.

Technology transfer is frequently perceived as being relevant only to industrializing countries. This paper shows its relevance to industry on a global scale, in terms of the broad objective of sustainable development.

Conclusion

This paper suggests that the concept of environmental sustainability can be made operational if governments set measurable policy targets and design policy mechanisms that support implementation. The new regulatory principle — pollution prevention pays — aims to promote competitive and environmentally sustainable industrial development. The requirement that pollution be reduced at source implies a requirement for technical or organizational change, or both, in the production process. This, in turn, requires that firms develop new technological and managerial capabilities, technological alliances with equipment suppliers, and collaboration with R&D organizations. This paper argues that for successful implementation of pollution prevention, regulatory approaches must be underpinned by technology-policy mechanisms designed to stimulate technological innovation and best practice in environmental management within firms and to encourage the commercialization and diffusion of these innovations across the boundaries of firms and nations. As a contribution to sustainable development, pollution-prevention policy represents an advance over previous policies, such as those guided by the polluter-pays principle. However, it contains two flaws.

First, the firms that pollute the most are mismanaging the environment precisely because of their inability to innovate. Environmental degradation is greatest in operations with low levels of productivity, obsolete technology, limited capital, and poor human-resource management. Yet, under the pollution-prevention regulatory regime, it is assumed that such firms, if obliged by law, will automatically introduce technical change to reduce pollution at source. This is unlikely to occur unless pollution-prevention regulations are underpinned by technology policies and financial incentives aimed at encouraging the least-efficient firms to develop the technological and managerial capabilities to innovate. Using punitive environmental regulations to put the least-efficient and most-polluting firms out of business is a short-sighted alternative. In developing countries particularly, such an approach would threaten the economic objectives of sustainable development and lead to further problems because the cleanup and mine-rehabilitation costs would be transferred to the state or society.

Second, the most efficient firms are generally better environmental managers because they are innovators. They are able to harness both technological and organizational change to reduce the production and environmental costs of their operations. Furthermore, where the costs of complying with environmental regulations threaten competitiveness, the dynamic firm can offset these costs by improving production efficiency. In the minerals industry, regulatory costs cannot be passed on to consumers because international metal prices are determined in terminal auction markets and cannot be controlled by the producers. The policy of requiring firms to reduce pollution at source, which necessarily involves changing their production technology and organization, overlooks the possibility that firms might already be searching for new ways to improve metal recovery, reagent use, energy efficiency, water conservation, and so on as part of their corporate strategies to increase competitiveness.

It is therefore more likely that pollution-prevention regulations will serve their objectives if they are underpinned by technology-policy mechanisms and economic instruments. The following approaches are suggested:

- Stimulate and reward innovation in pollution prevention through tax breaks for R&D and technology investment, other taxation reforms, auctioned pollution permits, new lines of credit, targeted R&D support, and training programs;

- Require mandatory pollution-prevention and reclamation plans in project development, and stipulate bonds for that purpose;

- Stimulate profitable innovation in waste management, such as re-mining, reagent and metals recovery, and biotechnological waste treatment, and remove legislative barriers to re-mining and waste treatment;

- Reward firms for innovations in clean technology;

- Use mechanisms such as credit conditionality to facilitate the commercialization and diffusion of pollution-prevention technology and work practices across the boundaries of firms and nations; and

- Promote new approaches to technology transfer, such as interfirm collaboration to develop the technological and managerial capabilities to innovate, in-depth training to manage technical and organizational change, and information-dissemination programs.

Innovation can change the context of metals production and pollution, and the widespread diffusion of innovation can reward the innovator, as well as contributing to best practice in environmental management for sustainable development. Mechanisms to support pollution-prevention policy will be more successful if they focus on the process of innovation at any point in the life cycle of the mine, rather than penalizing firms for excessive use of inputs or production of polluted outputs. The production of outputs varies too much among operations because of site-specific geological and geographic conditions. Penalties can differentially distort the operations' cost structures and be an inefficient way to stimulate innovation in pollution prevention.

This paper also makes a case for training regulators so that they have the experience and understanding to evaluate technological advance, an important indicator of the effectiveness of environmental regulations. Ratcheting the existing regulations in line with this evaluation would further enhance the competitive advantages of firms. Regulators and corporate analysts might also enhance their strategies for competitive environmental best practice by defining corporate environmental trajectories in various economic and regulatory contexts. This would help in evaluating the evolution of a firm's competitiveness in response to changing market conditions and regulatory requirements and, therefore, in evaluating the firm's contribution to sustainable development.

Broadening the range of regulatory goals and the technology-policy mechanisms and economic instruments to support them, as proposed here, would be a more integrated policy approach to regulating and promoting industrial development and to promoting trade and technical assistance abroad. Pollution prevention at source would have a key role in this policy, without always taking priority in the competitive and environmentally sustainable development of industry in developing and industrialized countries. This aim of this new, more comprehensive policy approach is environmental innovation.

CHAPTER 2[1]

US ENVIRONMENTAL REGULATIONS AND THE MINING INDUSTRY: LESSONS FOR CHILE

Juanita Gana

In the design and implementation of environmental regulations, the United States is far ahead of many other countries. With more than 20 years of systematic efforts to protect the environment and improve the quality of life for its citizens, the United States can offer several lessons to Chile and other countries with less experience.

For any country planning to pursue environmental stewardship, aspects of the utmost relevance are

- The development of environmental consciousness, the mechanisms of social pressure, and the policy-making response;

- The approach selected to deal with the problems, the tools used, and the ability of these to solve the problems with minimum adverse economic impact;

- The impact of environmental regulations on productivity, market structure, and economic growth;

- Regional impacts and their effect on employment; and

- The reaction of industry and the labour unions.

---

[1] The preparation of this report was made possible by a grant from the John D. and Catherine T. MacArthur Foundation. The author would like to acknowledge the contribution and support of all the interviewees. As always, the opinions expressed are solely the responsibility of the author.

It would not be prudent for other countries to simply duplicate either the style or the specific mechanisms of a specific US policy-making experience. For countries with different political, social, economic, and, especially, ecological conditions, it might even be dangerous. The lessons from the US experience, whatever they may be, must be adapted to the context in which they are to be used.

Also, we cannot say that the United States has found the ultimate formula for dealing with environmental problems. After two decades, the principal moral is that no such formula exists and that environmental policies should, above all, be flexible. The time may be too short, and we are still in the experimentation stage. The very nature of the problem forces us to recognize a deep void in our scientific knowledge about ecosystems and the impacts of human activities — a void that might never be filled to the extent that we feel secure about the consequences of our decisions. That has been the ultimate challenge of environmental problems: the increasing awareness of our poor understanding and our lack of control over nature.

The focus of this research on mining reflects the importance of this sector to the Chilean economy. This makes it of special relevance to identify the problems this sector poses for the environment, as well as become aware of the possible consequences that environmental regulations would have for this sector. These concerns are covered, and the present regulatory scheme and its impacts on the mining sector, particularly the copper industry, are also examined. The paper also discusses the efficiency, cost-effectiveness, dynamic efficiency, and equity of existing policies and mechanisms.

As mentioned, environmental policy in the United States is still in its trial-and-error phase. This paper examines new trends in environmental policy-making and the ways they might affect the mining industry. I take a closer look at the case of mining-waste disposal, which is receiving a lot of attention from the industry, environmental groups, and the government. The discussion of mining-disposal regulations involves not only the next step in the control of the industry but also an interesting experiment. New procedures and new concepts are being tested; their success may bring about important changes in ways of writing environmental policy. Finally, I summarize the main conclusions from the US experience and make some recommendations for policy formulation and implementation in Chile.

This study is by no means exhaustive: it is the product of a 3-month project and focuses on just some of the several themes relevant to environmental policies in the United States. Even then, the treatment of themes cannot help being somewhat superficial. Nevertheless, the study develops a sense of the main policy issues and establishes some guidelines for future policy-making.

US environmental regulations and the mining industry

The first US pollution-abatement regulations date from mid-century. The *Water Pollution Control Act* of 1948 and the *Air Pollution Control Act* of 1955 were enacted by Congress to address the increasing health hazards posed by industrial activities and the lifestyle of US society. The purpose of enacting these laws was mainly to grant the federal government the authority to allocate resources to investigate the causes and effects of pollution and to train human resources from state and local agencies. These laws also transferred some responsibilities from the state to the federal level. However, the state governments still had the authority to implement and enforce regulatory programs.

The need for a national framework and a stronger federal presence became more and more evident as environmental problems grew and public opinion became more sensitive. State legislation was dispersed and became a potential source of competitive disadvantage. As a consequence, states were often reluctant to take the initiative. Reacting to a strong environmental movement, Congress passed the *National Environmental Protection Act* (NEPA) in 1969, which was to become one of the most influential environmental regulations in the United States and abroad (Anderson et al. 1984).

NEPA was the first attempt to give a systematic and coherent framework to the problem, and it established a conceptual basis upon which other legislation was created or amended. Although NEPA gave little guidance on how its general objectives were to be met, it established a powerful mechanism for introducing environmental considerations into the decision-making process. This mechanism was the environmental-impact assessment (EIA), which NEPA required before any major federal action that would significantly affect the quality of the human environment could be undertaken. The EIA process forced federal agencies to take environmental concerns into consideration during the planning process. NEPA also made it possible to challenge federal actions affecting environmental quality, resulting in a number of high-profile court cases that served to raise public awareness and concern about environmental problems (Anderson et al. 1984).

The institution to implement NEPA was created in 1970. Several government agencies were already in charge of implementing and enforcing the several dispersed laws that in some way or another protected the environment, but Congress decided to create a new, separate agency, the Environmental Protection Agency (EPA). The rationale for the EPA was to fulfil the need for an independent institution with the expertise to formulate environmental regulations and to oversee their implementation and enforcement. Having a separate agency raised the issues not only of coordination and regulatory consistency and coherence but also of autonomy. Other government agencies were in charge of fulfilling several other objectives, with the environment being only one of them and probably not the most relevant (see, for example, the case of the US Atomic Energy Commission in Anderson et al. 1984). After this basis for environmental policy-making was set, frantic activity began in Congress.

As a consequence of its rather low national profile, mining pollution occupied a secondary place on the US environmental agenda during the last decade. The pollution produced by the chemical and petroleum industries seemed far more worrisome. But the environmental impacts of mining range from land disturbance produced by exploration, development, and mining activities, especially in the case of open-pit mining; to the pollution of surface water and groundwater by metals, toxic chemicals, and acid mine drainage; to the pollution of air by SO_2 emissions and the like. Fugitive dust may also be an environmental hazard, although its impact is mostly impaired visibility. Mining pollution tends to be very localized, and because the population is generally sparse around mines, fewer people are exposed to health risks and aesthetic effects than is the case with industrial pollution in suburban areas. Nevertheless, mining pollution may have important ecological and aesthetic effects (Gomez et al. 1979; Vogely 1985; MacDonnell 1989).

Regulations affecting mining were introduced because of broader concerns, with the result that the role of the mining industry in the policy-making process has been minor. The mining industry's loss of importance in the US economy and its diminished strategic significance have further reduced the industry's negotiating power. The fact that mining has not played an important role in environmental policy-making contrasts sharply with the impact that environmental regulations have had on the industry.

The relevant regulations and legislation include NEPA, the *Clean Air Act* (CAA), *the Resource Conservation and Recovery Act* (RCRA), *the Comprehensive Environmental Response, Compensation, and Liability Act* (CERCLA), *the Surface Mining Control and Reclamation Act* (SMCRA), and the *Mining Law*.

The impact of environmental regulations on copper mining

Much controversy surrounds the environmental regulatory framework and the burden it has imposed on the copper mining industry. Complaints about exaggerated costs and the loss of competitiveness have been recurrent. Negative impacts on employment and on regional economic activity have also been a part of the discussion.

According to the US Department of Commerce and its Bureau of Analysis (USDC 1988), expenditures for pollution abatement and control have been constantly rising since the early 1970s, except during the 1980–82 recession (see Table 1).

Table 1. Total expenditures for pollution abatement and control, 1972–87.

| | Total expenditures (billions of 1982 USD) | | | | | | | | | | | | | | | |
	1972	1973	1974	1975	1976	1977	1978	1979	1980	1981	1982	1983	1984	1985	1986	1987
Abatement costs	40	46	47	50	53	55	58	60	58	57	55	56	61	65	68	68
Air	15	18	18	21	22	23	24	24	25	26	25	26	28	30	31	28
Water	20	21	21	23	24	25	27	26	25	22	21	21	23	25	26	28
Solid	7	8	9	8	8	9	9	10	11	11	10	19	11	11	12	13
Total abatement and control costs [a]	43	49	50	54	56	59	62	63	62	60	58	60	64	68	72	71

Source: USDC (1988).
Note: USD, United States dollars
[a] Includes regulation and monitoring costs, as well as research and development expenditures.

Overall, these expenditures grew at an average annual rate of 3.4% during 1972–87. Currently, the annual level of expenses is close to 70 billion United States dollars (USD), about 2% of the US gross national product.

Among laws and regulations, the CAA has been considered by far the most expensive. More than half of the expense of air-pollution abatement is for controlling pollution from mobile sources, reflecting the importance of vehicle emissions as a source of air pollutants.

Another source, the McGraw-Hill (1982) annual survey, indicated that pollution-control expenditures on average accounted for more than 5% of total capital expenditures during 1975–79; 3%, during 1980–84. In the case of mining, McGraw-Hill estimated a total of 21.8 billion USD in capital expenditures for 1970–81. An EPA study cited in MacDonnell (1989) gave a significantly lower figure: about 8.9 billion USD. The EPA study also gave an estimate for 1981–90 of 5.3 billion USD. EPA used engineering estimates of costs for compliance with federal air and water regulations, whereas the US Department of Commerce and McGraw-Hill relied on industry surveys and may have included other regulatory costs.

In terms of total costs, including control and maintenance, the EPA study indicated a cumulative annualized cost of 15.5 billion USD for 1970–81. The annualized cost for 1981 totalled 2.6 billion USD, and the cumulative annualized costs projected for 1981–90 totalled 32.7 billion USD. Following the general pattern, the CAA has been the most expensive regulation for the mining industry (SMCRA in the case of coal production). According to the same EPA study, cited in MacDonnell (1989), about 80% of the mining industry's expenditures for pollution abatement in 1970–81 were for control of air pollution; the other 20% were for control of water pollution.

In absolute terms, the iron and steel industry has been the most affected by air- and water-pollution-control costs, followed by the copper industry. Of the total 8.9 billion USD of investment reported by the EPA, 4.6 billion USD was spent by the iron and steel industry, and around 2.1 billion USD was spent by the primary copper industry (the last figure includes only costs for air-pollution control). Nevertheless, in terms of pollution-abatement costs as a proportion of total capital expenditures, the copper industry shows an outstanding 41%, whereas the iron and steel industry shows only 18% (Sousa 1981).

Because copper is the most important mineral in Chile, the following pages concentrate on the impact of the CAA on the copper-smelting industry. In the United States, this industry has been one of the sectors most affected by environmental regulations, particularly the CAA (Sousa 1981).

The CAA and the copper-smelting industry

The CAA regulates SO_2 emissions and sets primary and secondary standards for this pollutant. Sulfur dioxide is a primary focus of the CAA. The pollutant has a number of negative impacts, including aggravation of symptoms of heart and lung disease and increased incidence of acute respiratory disease. It can also be toxic to plants, erode statues, corrode metals, harm textiles, impair visibility, and contribute to acid deposition (GAO 1986).

The principal sources of copper are sulfide deposits. The production of each tonne of copper releases an equal or greater volume of sulfur. The recovery of copper from sulfide ores is done by pyrometallurgical processes that separate the copper from other elements like sulfur. After the copper ores are ground and concentrated, the concentrate is smelted, passing through a circuit of furnaces, converters, and roasters, any one of which may release sulfur into the atmosphere as SO_2. Finally, the blister obtained from the smelters is refined. For a description of these processes, see Rothfeld and Towle (1989).

Oxide ores used in the production of copper do not present this SO_2 problem because they are treated by hydrometallurgical processes. Unfortunately, oxide ores play a minor role in the copper industry because of their relative scarcity — only about 16% of US copper production involves oxide ores (Rothfeld and Towle 1989). However, they present higher risks of water pollution (USDC 1979).

At the national level, the major source of SO_2 emissions is the power-utility industry (Table 2). Copper smelters make a significant contribution, especially at the regional level. Both industries together generate more than 70% of total SO_2 emissions in the United States (Rothfeld and Towle 1989). According to EPA figures cited by the US Bureau of Mines (USBM), utility boilers generated 14.7×10^6 t of SO_2 in 1985, whereas copper smelters produced only 0.6×10^6 t. But in the states with the greatest smelting capacity — Arizona, New Mexico, and Utah — utility boilers generated 0.2×10^6 t of SO_2, whereas copper smelters generated 0.6×10^6 t of SO_2 (USBM 1989) (Table 3).

Table 2. Sources of SO_2 emissions in the United States, 1980.

Source	SO_2 emissions	
	$(\times 10^6$ t)	(%(
Utilities	15.8	62.7
Nonferrous-metals smelters	1.4	5.5
Copper smelters	1.1	4.4
Others	6.9	27.4

Source: GAO (1986).

Table 3. Regional distribution of SO$_2$ emissions from power utilities and copper smelters, 1985.

State	Copper smelters (× 10^3 t)	Utility boilers (× 10^3 t)
Arizona	454.5	106.9
New Mexico	96.6	70.7
Utah	8.3	22.1
Total for 3 states	559.4	199.7
Total for 48 states	578.0	15 249.9

Source: Rothfeld and Towle (1989).

Yet these SO$_2$ emissions levels represent an important CAA accomplishment, because copper smelters played a much more important role in SO$_2$ emissions a decade ago. In 1980–88, SO$_2$ emissions from copper smelters were reduced by 73%, from 1.1×10^6 t to 0.3×10^6 t. In 1987, the aggregate capture of SO$_2$ emissions was 83% (Rothfeld and Towle 1989). Control is currently estimated at far more than 90%, thanks to the retrofit of San Manuel (Magma Copper Company) and additional improvements derived from previously retrofitted plants. With currently available technology, it is possible to capture 99% of SO$_2$ from the gases released from smelters (Rothfeld and Towle 1989).

Before air-quality standards and emission limitations were enforced at the federal level, some degree of control was provided at the state level. Some smelters had already decided to recover SO$_2$ to produce sulfuric acid, even without regulations, as in the case of Garfield Refining Company. In such cases, around 60% of SO$_2$ emission was captured from roaster and converter gases. The remaining 40% came basically from reverberatory-furnace stacks and from fugitive gases from what were mostly old plants. The problem with recovering SO$_2$ emissions from reverberatory furnaces — a technology widely used some decades ago with copper smelters — was the weak gas stream, which contained less than 1% of the SO$_2$. It was technologically and economically impracticable to recover SO$_2$ under those conditions, even though there was a possibility that some technical obstacles could be overcome (Rieber 1986; Rothfeld and Towle 1989).

After implementation of the 1970 CAA, some smelters installed acid plants and used tall smokestacks and intermittent controls to comply with the ambient-air-quality standards. But the 1977 amendments to the CAA prohibited the use of these techniques for stationary sources and required permanent controls. The only alternative for copper smelters was to replace the reverberatory furnace with other technologies, like flash or electric furnaces or bath smelting. (The use of scrubbers, a technology that power plants use to remove the sulfur from coal, is not economically feasible for copper smelting because of the greater amounts of sulfur involved. There are also additional storage costs and so on [Rieber 1986].) These other furnaces and the use of enriched oxygen did provide a gas stream strong enough to allow the recovery of SO_2 (Sousa 1981; Rothfeld and Towle 1989).

The huge investments needed to retrofit the plants, most of them built in the early 20th century, led the industry to ask for some relief. In addition to the financial burden, the industry cited a lack of proven technologies for controlling high levels of SO_2 emissions. Eventually, the industry did get some relief — the Non-ferrous Smelter Orders (NSOs) provision. The NSOs allowed smelters to delay new investments and to use temporary measures, like curtailed production or taller stacks, to comply with ambient-air-quality standards. The NSOs extended compliance deadlines by 5 years, with the possibility of a second extension.

Although the provision was intended to apply to all nonferrous-metals smelters, only copper smelters — San Manuel (Magma Copper) and Douglas (Phelps Dodge), in Arizona; Chino (Phelps Dodge [by that time owned by Kennecott Corporation]), in New Mexico; and McGill (Kennecott), the only Nevada smelter— requested NSOs. After these NSOs expired, San Manuel, Douglas, and Chino requested a second round (McGill had shut down in 1983). Although San Manuel obtained a second NSO, Douglas shut down in 1987. Chino was retrofitted before the final decision. The government also gave the industry some financial support by allowing rapid amortization of pollution-control equipment and by providing tax credits for such investments (Larsen 1981).

The first smelter to change its process technology was Inspiration, which converted to electric furnace in 1974. The last smelter to invest in SO_2 control will be El Paso, owned by Asarco Incorporated. In the last three decades, there has been only one new greenfield project, Hidalgo, owned by Phelps Dodge. Hidalgo began operating in 1976. From the beginning, it introduced air-pollution-control technology, and it was, at the time of its construction, considered to be the most modern and efficient copper smelter in the country (Rothfeld and Towle) 1989.

Technology

EPA standards demand a permanent end-of-pipe type of control. Dispersion techniques have been explicitly prohibited, as well as temporary reductions in production levels. Consequently, legislation has implicitly imposed the SO_2-fixation method for reducing SO_2 emissions. This has meant replacing old reverberatory furnaces with other equipment to recover SO_2 and produce sulfuric acid.

The US smelting industry uses flash and electric furnaces, but bath smelting has also been used in other countries. Other techniques are either unproven for use at an industrial scale, like the ammonia scrubbing system, or too expensive to use with copper concentrates, like the limestone scrubbing system used in coal-fired electric-power plants.

For the most part, smelters choose the technology that suits their own site-specific conditions, such as metallurgical parameters or input supplies. The tendency has been to use Outokumpu smelting technology, which accounts for more than 75% of the flash furnaces currently operating in the world and two-thirds of the new capacity. Although the Outokumpu technology has important advantages, such as being the lowest-risk option, the Inco and Noranda processes also have advantages, such as simplicity and the capacity for handling dirtier concentrates. Each of these technologies achieves full compliance with strict environmental standards, but the Mitsubishi continuous smelting process yields the highest level of SO_2 fixation (more than 99%).

Although the replacement of reverberatory furnaces has brought additional capacity, increased productivity, and energy savings, these investments might not have been made had there been no regulatory requirements (Sousa 1981; Rieber 1986; Cook 1989; Roethfeld 1989). The new technologies reduce operational costs, but in terms of capital costs, the scale shifts in favour of the old technology (Cook 1989). For greenfield projects, however, the new technologies have smaller capital and operational costs than a reverberatory furnace does, according to Burckle and Worrell (1981).

A company's decision to produce sulfuric acid is also considered a consequence of the regulatory environment, because the sulfuric acid market by itself fails to justify its production (Rieber 1986; Rothfeld and Towle 1989). Nevertheless, some smelters have long been producing sulfuric acid, like the Garfield smelter, which built an acid plant in 1916 (Navin 1978). In any case, ore leaching and electrowinning are creating an interesting alternative market for sulfuric acid.

The introduction of these technologies was not trouble free. Although they were being used in other countries, they had not been tried at full scale or under the metallurgical conditions of the US smelting industry. Temporary closures, delayed start-ups, and productivity losses were part of the costs of complying with the regulations. Moreover, after the new technologies were introduced, some smelters still had problems complying with emission standards — Inspiration and Hayden (Asarco) are examples (GAO 1986).

The cost to the industry

Analyzing the costs that environmental regulations imposed on the copper industry, the USBM (1989) determined that the principal impact was on smelting, because compliance with the CAA entailed major process changes, substantial capacity reduction, and increasing export of ores and concentrate. Other researchers drew the same conclusion. According to a study done for EPA in 1978 by A.D. Little, Inc. (Sousa 1981), 24% of the copper industry's total investment in the 1972–75 period went to pollution control. For copper smelters, the figure was 74%. Only 4% of this investment was for water-pollution control; the other 96% went to air-pollution control.

The strong impact of the CAA and its SO_2 standards on the copper industry stimulated several studies. Some of these studies were prepared for the EPA, some were prepared for the industry, and at least a couple were prepared for Congress to use in considering protection for the domestic copper industry. USBM's Minerals Availability Program prepared the most complete report. This study, by Rothfeld and Towle (1989), examined the remaining seven southwestern smelters (which accounted for 96% of the US smelting capacity in 1987) and identified the regulatory impacts, including monitoring and direct administrative costs. The study concluded that, on average, environmental, health, and safety regulations added 0.032 USD/lb (1 lb = 0.454 kg) to operating costs. Sulfuric acid credits reduced this figure to 0.019 USD/lb. (For comparison, the total operating cost for an average smelter was 0.123 USD/lb — see Table 4.) These regulations also added 0.031–0.104 USD/lb to capital costs, depending on the smelter. (These calculations assumed operation at full capacity and excluded administrative overhead and indirect costs. The calculation of capital costs assumed a 15% rate of discount.) Metallurgical conditions, size of the plant, degree of obsolescence, and technological choices were the main factors affecting compliance costs. If both operational and capital costs are taken into account, compliance with environmental regulations represented 45% of total smelting costs and 14% of the total costs of producing a pound of refined copper (assuming no capital costs other than regulatory capital costs). Rothfeld and Towle also indicated that the capital costs of retrofitting a smelter averaged 150 million USD.

Table 4. Operational costs for an average smelter.

Source of costs	USD/lb	%
Labour	0.0465	37.8
Energy	0.0374	30.4
Supplies	0.0391	31.8
Total cost	0.1230	100.0

Source: Rothfeld and Towle (1989).
Note: USD, United States dollar; 1 lb = 0.454 kg.

The calculations given by Rothfeld and Towle (1989) take into account productivity gains resulting from new and improved technology, lower energy costs, and greater production capacity. These calculations also include health and safety expenses, but according to a study cited in Sousa (1981), 95% of the total regulatory expenses are attributable to compliance with EPA standards. On the other hand, those figures may be considered conservative because they include only direct costs and do not take into account the opportunity costs involved in the slow process of obtaining permits — legal fees, red tape, and delays add to the costs.

Several other studies tried to measure the actual costs of environmental regulations or to assess possible impacts of full compliance and stricter standards. For example, the Congressional Research Service (CRS 1984a), analyzing different sources of data, gave a cost for full compliance — not necessarily effective compliance — ranging from 0.05 to 0.15 USD/lb. An industry source gave the highest estimate, but the State of Arizona gave the most probable estimate, an average of 0.09 USD/lb. Earlier studies, like Sousa (1981), gave effective costs ranging from 0.03 to 0.05 USD/lb and projected an additional cost of 0.10 USD/lb for full compliance with the 90% emission-control standard.

Studies tended to overestimate future environmental costs because they normally included the cost of compliance for smelters that would shut down. These smelters usually had the highest retrofitting costs. Exploratory studies also failed to take into account substitution effects or technological improvements. Although actual costs were less than previously estimated, the relative impact of environmental regulations on operational costs was greater, because operational costs were reduced through modernization.

It is also interesting to notice that the estimates prepared by governmental agencies at the beginning of the environmental era often underestimated the real costs of compliance by the copper industry. For example, in 1971 the President's Council on Environmental Quality (CEQ) estimated that air- and water-pollution control would require capital expenditures of 311–682 million USD (MacDonnell 1989). If operation and maintenance are included, the cost goes up to 346–758 million USD (Charles River Associates Inc. 1971). Even if we take the highest point of the CEQ estimate and adjust for inflation, actual costs were more than double the estimated costs.

This underestimation of real costs was rather common. A lack of experience and a poor understanding of some industries and their technological challenges most often led to optimistic assessments of the economic impact of regulations. Certainly, in the case of the copper industry, the dramatic changes that took place in the US economy and international markets did not help.

If projecting environmental costs proved to be a difficult task, identifying the actual cost of compliance for the smelting industry was not much easier, as we have already seen. The current accounting system makes no clear distinctions among regulatory costs, leading sometimes to important discrepancies, depending on the methodology used for estimating. Figures given by companies usually refer to total investments, without adjustments for increased productivity or higher energy efficiency; some may be exaggerated just to improve the external image of the company (Gulley and Macy 1985).

The CAA amendments of 1990

The CAA amendments of 1990 introduced additional controls for SO_2 emissions and focused on the power-utility industry. Nevertheless, the regulations pertaining to toxic substances released into the air are a possible new source of compliance costs for copper smelters and for the copper industry in general. The Bush administration estimated the annual cost at 3 billion USD to the whole economy. Industry estimates ranged from 14 billion to 62 billion USD (Portney 1990). Uncertainty about the cost to the mining industry is even greater.

Production and employment levels

Probably the most dramatic impact of environmental regulations has been the shutdown of several smelters (which may be evidence of an overwhelming financial burden) and, as a consequence, the reduction of the national smelting capacity. An equally visible consequence has been the reduced levels of employment in the industry.

In 1970, the United States had 17 smelters, and the total primary smelting production was about 1.6×10^6 t of copper. Two decades later, in 1989, the number of smelters had been reduced to eight, and production had been reduced to 1.5×10^6 t of copper; there was one new greenfield project. Smelting production reached its lowest level in 1983, with 1×10^6 t of blister and anodes. The smelting and refining industry suffered a steeper decrease in employment, from an estimated 11 600 workers in 1967 (Charles River Associates Inc. 1971) to about 5 400 workers in 1988 (USBM 1989).

The reduction of capacity and production has certainly been an important factor in accomplishing environmental goals; such reductions may even be an inevitable short-term consequence of implementing environmental controls. According to GAO (1986), 56% of the reduction in SO_2 emissions from nonferrous smelters was achieved because of reduced production; only 44%, because of retrofitting and new technologies.

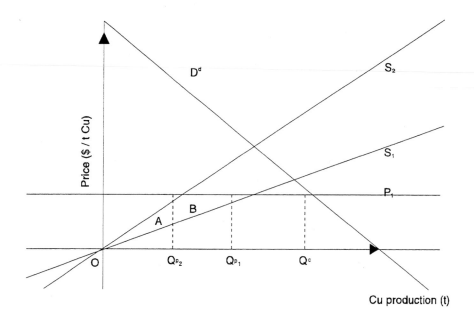

Figure 1. Cost of copper production.

The shrinkage of the smelting industry — and of the copper industry in general — is certainly an additional cost to society. This can be seen in Figure 1, where area A represents the higher operational costs, already discussed, and triangle B represents the cost of reduced production and employment (the smaller the mobility of capital and human resources, the larger the triangle).

This cost has not been included in most evaluations of the impact of pollution controls on the industry — estimates have concentrated on operational costs. Nevertheless, attempts have been made to estimate the impact on total capacity and employment; at first, the estimates were far too low. CEQ, in its "most extreme scenario," projected the stabilization of smelting capacity at around 1.6×10^6 t. It also fell short in employment estimates, predicting that employment would not fall below the 1970 level. In 1970, total employment in the copper industry was estimated to be 54 000; by the end of the next decade, it was around 18 000 (MacDonnell 1989).

If we use the triangles in Figure 1 and apply the capacity lost during these last decades and Rothfeld and Towle's (1989) operational costs of compliance, we obtain a figure close to 63 million USD. This may be an underestimation, as the calculation of regulatory costs took into account only direct costs, and we are not allowing for any expansion and are assuming perfect mobility of resources. But it may also be an overestimation, as the enforcement of environmental regulations was not the only reason for the reduction of US copper-smelting capacity. As we review the circumstances, we find that some shutdowns would have taken place regardless of environmental regulations (Mikesell 1988). In fact, even some plants that were already in compliance and thus not threatened by new capital expenditures closed. In fact, 5 of the 10 closed smelters already had control equipment in place to meet SO_2 emission limits (GAO 1986).

Several factors compounded the difficulties for the copper industry during the late 1970s and early 1980s. The most important was the crisis in the international copper market (Mikesell 1987). Copper prices reached their lowest levels since the early 1930s. Excess capacity (created during the early 1970s in an over-optimistic reaction to good market conditions) and declining rate of growth in copper consumption caused the glut in the market. Lower-quality ores, higher labour costs, and older plants made it difficult for US copper producers to confront a more competitive market. In addition, some companies were going through hard financial times.

International competitiveness

Environmental regulations have corroded the competitiveness of US copper producers and reduced their world-market shares. In the US market, increased imports of refined copper have compensated for decreased domestic production. This effect does not necessarily increase the social cost of environmental regulations unless we consider, first, that there is a premium for reduced vulnerability and, second, that the United States is not a marginal actor in the international market.

If we look at the first consideration, we find that although some decades ago copper was a strategic material, times have changed. Nowadays, copper is a traditional metal with many possible substitutes, and external supplies come from allied countries that are fairly stable politically and economically. In 1986, the worst year of the crisis in the copper market, refined-copper imports reached a maximum of 23.5% of the total apparent consumption (Mikesell 1987).

Regarding the second consideration, we can say that although the United States is one of the most important copper consumers and producers in the world, its role in the international market is not decisive. In 1986, it imported slightly more than 500 000 t of refined copper. Moreover, most of this trade was with Canada, to some extent a captive market.

The industry argued that environmental regulations were one of the main reasons it needed protection against copper imports. The additional costs imposed by compliance with air-pollution and other controls were identified as a significant factor in the domestic industry's loss of competitiveness. Other producing countries without similar standards and requirements were said to be subsidized.

Several studies were done to assess the real damage and the need for relief. As mentioned earlier, this concern stimulated most of the research on the costs of environmental regulations to the copper industry. Congress discussed the possibility of protection for the domestic copper industry in 1978 and 1984. In both cases, although the International Trade Commission recommended import relief under the escape clause of the *Trade Act* of 1974, Congress ultimately denied the copper industry's petition for import protection.

The crisis affecting the US copper industry was not unique. Other base-metals industries were experiencing similar disruptions, as was the iron and steel industry. The issues were similar: depressed markets, loss of competitiveness, temporary and permanent closures, and unemployment. Environmental regulations, particularly air-pollution standards, were also an issue, although not as much as in the copper industry (Crandall 1981).

The structural causes of the crises were also similar. Although environmental controls were imposing a significant burden on US industry, other factors played a role. The most important of these was the change in the traditional behaviour of the international metals markets. The growth rate of world demand for all basic metals was decreasing; to an important extent, this was a result of the energy crisis and the increasing concern about materials-use efficiency. The economic recessions of the mid-1970s and early 1980s somewhat obscured the reduction in the materials-use intensity indices (Tilton 1990).

On the supply side, equally important changes were taking place in the late 1960s with the emergence of new low-cost producers, who increased competition in international markets. In the copper industry, an important change was the nationalization process in the late 1960s and early 1970s, which not only altered the organization of the industry and augmented the role of state enterprises (Sousa 1981; Cook 1989) but also affected companies with incomes heavily dependent on their *filials* (subsidiaries). Direct investment losses were equally significant for some of them, such as Anaconda (Navin 1978).

US industry was badly prepared for the new scenario. Obsolescence was a problem for most of the plants, especially compared to the state-of-the-art plants being built in countries like Japan and Korea (see, for example, Adams [1986] for the case of the iron and steel industry; Sousa [1981], the copper industry). An equally relevant factor was higher labour costs.

Productivity growth in the copper industry decreased during the 1970s and early 1980s, but wages continued to rise at the normal rates. For copper smelters and refineries, Sousa (1981) found a negative rate of productivity growth (averaging −0.3%) for the 1960s and 1970s.

High interest rates in the United States also eroded the position of these US industries, which needed to undertake big modernization projects to comply with environmental regulations and deal with the competitiveness crisis. Some companies were already highly indebted and burdened financially (Navin 1978; USDC 1979), and the appreciation of the dollar exacerbated these problems (Sousa 1981; CRS 1984). For the copper industry, a more fundamental factor was its reliance on ores of relatively low quality (Sousa 1981; CRS 1984).

One of the reports prepared for Congress concluded that environmental costs were one among several factors affecting the industry (CRS 1984b). After balancing the different factors generating the differential between the costs borne by US producers and those borne by their lowest-cost competitors, the report suggested that import protection would give only temporary and marginal relief and would not address the root problems of the industry.

Although Congress was concerned about the economic and social impacts of additional shutdowns in the copper industry, other considerations also influenced its final decision to deny the petition for protection. One consideration was that such protection might adversely affect the copper-user industry, possibly shifting the competitiveness problems to that sector. Other considerations were the possibility of complaints under the General Agreement on Tariffs and Trade and other treaty obligations and the possibility of negative impacts on the economic stability of allied copper consumers (CRS 1984b).

Although the iron and steel industry got secured trigger prices during the 1970s and import quotas in the 1980s, the copper industry was less successful and had to overcome the crisis without special protection. Nevertheless, in the end, both had to accept drastic restructuring. High-cost plants and producers went out of business, and those who remained had to reorganize production, renegotiate labour contracts, and undertake major modernization projects. In the copper industry, the results were impressive — the average production cost went down 42.5% between 1981 and 1989, in real terms. Productivity increased both at the industry and at the smelter and refinery levels (USBM 1989).

Investment trends

Frequently, the industry has argued that the regulatory framework and its financial burden will lead to the migration of US investment to countries where regulations are less strict or nonexistent. Some analysts, like MacDonnell (1989), have supported this hypothesis, but the complexity of the subject makes it difficult to assess the real impact of environmental regulations on investment decisions.

The hypothesis and some caveats

The logic of the arguments for the investment-migration view, at first glance, seems to be crystal clear: in loosely legislated countries, mining production is less costly and more profitable. Available data on capital expenditures for pollution abatement confirm that US multinationals spend considerably less overseas than at home (UNCTC 1985).

But some considerations weigh against this view. One is that technology tends to be homogeneous, especially the processes and equipment used to produce internationally traded commodities. Companies investing overseas will use the technologies developed in countries already subject to strict environmental regulations.

Another important consideration is the reputation of the enterprise. Big multinationals with headquarters in developed countries would not like to be perceived as taking advantage of other countries and damaging their environment. The old stereotype of the multinational company stripping the assets of developing countries as fast as it can no longer pertains. In the late 1960s and early 1970s, the relationship between multinationals and developing countries changed dramatically. The multinationals have been more concerned to understand and fulfil the expectations of developing countries, creating a new type of partnership. Furthermore, stakeholders and environmental groups in home countries have begun to exert pressure on multinationals to protect the environment; sometimes this pressure is even greater than that exerted by host governments.

Although governments of some developing countries may have a more lenient attitude, the trend is toward increasing environmental control. Developing countries are introducing environmental concerns into their development projects, either voluntarily or because they are obligated to do so by the policies of international funding organizations, such as the World Bank and the Inter-American Development Bank. This trend toward increasing environmental control has in some cases created a disincentive for overseas investment. Lumpy and politicized processes add to the uneven capacity of these countries to formulate and enforce environmental regulations (Leonard 1988).

Nevertheless, to the extent that developing countries tend to use the experience of developed countries, there may be a trend toward a homogeneous treatment of environmental policies around the world. Unless their investments are short-sighted, multinationals will likely anticipate future changes and introduce environmentally friendly technologies from the start.

Yet, homogeneous treatment of environmental policies does not imply that the costs of compliance would be the same everywhere (Leonard 1988). The costs of compliance depend on regional environmental conditions and availability of inputs. Cost differentials may also result from building more-efficient institutional structures and less-cumbersome administrative procedures, and this is also possible in developing countries that have the will to learn from the experience of other countries.

It is not easy to discern from the available data the impacts of environmental regulations on investment in the mining industry. Certainly, increased costs have eliminated marginal projects in the United States, and even expansions have been affected to the extent that EPA's New Source Performance Standards (NSPS) also apply to new equipment. But as we have seen, many other factors affected firms' decisions in the 1970s and 1980s. International-market conditions were no incentive to invest anywhere, unless in highly profitable projects like small polymetallic deposits with high ore grades or expansions with low operating costs.

Political and general economic conditions added to the problems. During the 1970s, the relationship between multinationals and host countries suffered drastic changes, provoking the flight of mining multinationals and a general distrust of the stability of foreign-investment regulations in least-developed countries (LDCs). During the early 1980s, LDCs became less antagonistic as their external debt increased and international capital markets became more elusive. Some overseas investors benefited from the LDCs' critical need for capital and foreign resources. The trend toward the privatization of state companies made these countries attractive in the 1980s. Leonard (1988) had this to say about multinationals' overseas-investment decisions:

> When US companies, even those facing extreme pressure because of pollution problems at home, decide to build a plant abroad instead of in the United States, they do not necessarily do so because of differentials in pollution controls or because governmental and public concern for the environment may have delayed construction. Conversely, an industrializing country may have no intention of becoming a pollution haven, but other forces may induce it to attract certain high-pollution industries just the same. Thus, a major methodological problem is that it is difficult to single out the effects of any one factor in assessing either international comparative advantages or individual industrial-location decisions.

Leonard examined US Department of Commerce data on direct investment of US companies overseas in the late 1970s and early 1980s. He found that the mineral-processing industry's share of total US foreign direct investment did rise by a few points and that the portion of investment directed to developing countries also grew slightly, particularly during the early 1980s. An important portion of that investment went to Brazil and Mexico. The pattern is somewhat similar but more pronounced for the percentages of total capital expenditures going to LDCs. Leonard (1988) concluded that "stricter American environmental regulations have contributed to the international dispersion of some basic mineral-processing industries, such as copper, zinc and lead processing." Nevertheless, he added, "this trend is enhanced by other factors, such as the changing availability of raw materials, other nations' requirements that minerals be processed in the country where they are mined, and various economic factors including low prices, high interest rates, and recessions."

In the copper-smelting industry, only one greenfield project and some expansions were initiated during the 1970s and 1980s, but the situation now looks brighter. Plans for the 1990s include the expansion of the Cyprus smelter by 50% and the construction of a new smelter in Texas by Mitsubishi. Although environmental costs have affected the US industry, the location of the Mitsubishi smelter indicates that other factors have a more important influence on an investment decision. Mitsubishi is also involved in the biggest copper-mining project of the 1990s, La Escondida in Chile, but that project does not include smelting capacity.

Generally, as copper prices recovered, US production levels increased substantially. New projects are developing, which confirms that, overall, market conditions are still the main driving force. The improved market conditions will allow a better assessment of the impact of environmental regulations. Environmental regulations are the cause of cost differentials, but general wisdom and interviews with mining companies indicate that market considerations, the quality of the ore deposits, and long-term stability mostly guide their investment decisions.

Trends in policy-making and waste mining

The design, implementation, and results of a certain policy depend on more than economic factors: political and institutional considerations also come into play.

The setting for environmental policy in mining involves the role of mining in the US economy and, equally important, in the regional economy. A little bit of history and a look at recent structural changes help to explain the attitude of the industry, its involvement in the policy-making process, and its power of negotiation. Other elements are the structure of the industry and features of the international markets.

The reaction of the industry

Complaints were registered by the industry about the competitiveness of the eight plants closed in 1985; four had introduced technologies to control air pollution. White Pine was among those that reopened. So was Garfield, after a big modernization program that placed the plant among the lowest copper producers. Ray (Kennecott) and Ajo (Phelps Dodge) were closed for other reasons. McGill (Kennecott), despite having no pollution-control equipment, was not violating the EPA's National Ambient Air Quality Standards (NAAQS), according to the General Accounting Office (GAO 1986). According to Sousa (1981), it was. Although Kennecott rebuilt the reverberatory furnace, more stringent standards were introduced and it had to operate with NSOs. It was uneconomical to build a sulfuric acid recovery plant, and in 1981 it was evident that the plant was going to close (GAO 1986).

Of the plants that closed later, Tennessee Chemical apparently was in compliance; Douglas was not, and although it requested a second NSO, it looked like it was only a matter of time before it had to close (Rieber 1986):

> The closure of the Douglas smelter by 2 January 1988 is virtually assured, with or without a binational agreement. Given the plant layout, the age and type of furnaces, the projected state of the US copper market and the problem and costs of acid sale or disposal, Phelps Dodge will not build an acid plant. Given the first three factors alone, it is very doubtful that, even if an acid plant were emplaced, SO_2 capture would meet NSPS. Although its present smelter operating profits are favorable vis-à-vis other US smelters, this antiquated facility could not bear the financial burden of new equipment and APC [air-pollution control] in the existing plant.

As Sousa (1981) reminds us, it was uneconomical to built a sulfuric acid plant.

Morenci, the principal violator of NAAQS in Arizona, had already paid 682 500 USD in fines before it closed. Ajo and Hayden were also contributing and paid 25 000 and 52 500 USD, respectively. Phelps Dodge was planning to invest 195 million USD in its Morenci and Ajo operations and was negotiating with EPA in 1981. Ray was supposed to be redesigned to achieve 90% control in 1983. According to Sousa (1981), however, White Pine – Copper Range Company processed copper concentrates with a low sulfur content and was in compliance with the standards.

At first, the reaction of the industry was to resist the new regulations and to delay compliance. Citing considerations like a heavy financial burden, higher operational costs, and a lack of proven technology, they sought relief. To support their arguments with numbers, they hired various consulting firms to study the situation. They did achieve section 119 of the 1977 amendments.

The copper industry was not the only one having problems with compliance. A common feature of environmental policy-making has been the underestimation of the costs of compliance and, generally speaking, an overestimation of the technological capabilities of the industry. The 1977 amendments recognized how unrealistic the original deadlines were. "Despite the high cost and technical uncertainties implied in replacing all reverberatory furnaces by other smelting processes, according to the Arthur D. Little report economic considerations were absent in the establishment of NAAQS" (Sousa 1981).

But the copper industry got more than just a new deadline — it also got the NSO, a mechanism that was not available to all industries. In part, Congress seems to have considered the risk of closures and shutdowns resulting from a depressed copper market (GAO 1986). Even with the NSO, however, the industry was still having trouble complying with the NAAQS, and those companies not in compliance in 1977 did little to improve this situation. All of the nonferrous smelters requesting NSOs were copper smelters. Three of the four that obtained NSOs applied for a second period; the fourth simply went out of business.

The Division of Stationary Source Enforcement (DSSE) reported in 1978 that almost 50% of the 27 nonferrous smelters were operating in violation of the regulations governing SO_2 emissions. In contrast, by the end of 1979, "only 6.2% of the major air pollution facilities identified by the EPA were not in compliance with regulations." In 1980, DSSE reported litigation over the State Implementation Plans (SIPs) of Arizona, Utah, and Nevada. In the meantime, one smelter closed, and "the closing of this smelter has resulted in the only change in the compliance situation."

It is interesting to notice that more than half of the reduction in violations in 1977–86 was achieved by the closure of some smelters in 1984 and 1985. This is not surprising if we consider that only Douglas was releasing around 300 000 t of SO_2 a year. The report of the State of Arizona indicated that in 1984, the copper smelting industry incurred only one-third of the total cost of compliance.

The strategy of the states in developing their SIPs and enforcing compliance with the NAAQS depended to some extent on the industry's importance to the regional economy. Their approach was generally to negotiate first and to use court orders as a last resort; this was in part because of the huge legal expenses and time involved.

A comment on recycling

Arizona has had the highest levels of SO_2 emissions from copper smelters and the highest number of NAAQS violations (GAO 1986), partly because this state has more than 50% of the US smelting capacity and partly because its copper smelters have had more problems with compliance. Of the seven smelters operating in the 1970s, four have since shut down. This may have been one of the reasons Arizona's SIP approval took so long (Rieber 1986). Some sources indicated that other factors contributed to the plant closures. Sousa (1981) recorded that while US copper firms have relied largely on mining technology to maintain their competitive position in the world copper market, smelting technology in this country has not progressed at the same rate. Continued reliance on scale economies to reduce costs will likely yield diminishing returns.

To some extent, productivity growth did decrease because of the diminishing returns on the use of economies of scale (CRS 1984), not only in the smelter and refinery but also at the extractive level.

Innovation in the mining industry is likewise a difficult subject: recent studies found that only 10% of both federal and private investment in nonfuel-materials research and development (R&D) deals with minerals supply; the remaining 90% is directed to materials utilization. The studies also found that industry invests about four to five times as much as the federal government on R&D related to nonfuel-materials supply. However, the R&D intensity of non-ferrous metals industries is well below average.

Transboundary issues

The control of SO_2 emissions has had another kind of international dimension: SO_2 particles can travel many miles and generate acid rain — either wet or dry — far from the original source of the emissions. Consequently, flows of SO_2 emissions into and out of the United States have created conflicts with Canada and Mexico. But the flow to Canada is estimated to be more than double that from Canada. US SO_2 emissions appear to cause pollution problems in southeastern Canada. Most of these emissions come from power utilities and industries in the northeastern states. The main sources of Canadian SO_2 emissions are the nonferrous smelters in southeastern Canada.

Canada has been complaining for a long time about the problem, but in the 1980s, US policy and resources were focused mostly on studying it. Several meetings and special commissions were set up to study the issues, without specific outcomes. With the 1990 CAA amendments, however, the United States began to specifically address the problem.

There were several reasons for the delay. Most important were the economic impact and the equity issue for US power utilities and coal producers. Copper smelters played a secondary role because they were not the main offenders and the controls already in place made them a minor source of the problem at a national level. The US interest in a free-trade agreement with Canada may have been a major factor in the Bush administration's strong support for the initiative.

Similar problems have plagued the relationship between the United States and Mexico, although apparently these are not as salient as the US–Canada border issues. For a description of the transboundary issues in US–Mexico relations and the agreement to limit SO_2 emissions from the "Gray triangle," see Rieber (1986).

New trends in policy-making and mining-waste disposal

Environmental problems are complex and evolve with time and the social and economic contexts. Moreover, adjustments and sometimes major modifications in the system have to be made because of the relative lack of previous experience.

New knowledge about the health and environmental impacts of economic activities has stimulated policymakers to modify existing standards or create new ones. In the 1990s, the focus has been shifting to regional and global environmental problems, such as acid rain, ozone depletion, and global warming. Small and nonpoint pollution sources will become the targets of future efforts as the major polluting industries and firms are brought under control.

The approach to old and new problems will tend to be more integral, involving multimedia. Major efforts will be made to better prioritize environmental problems and to concentrate available resources on those issues with higher risks to human health and the environment.

Policymakers will be under greater pressures to change the policy mechanisms for dealing with environmental problems. Different sides are seriously criticizing command-and-control regulations because it is doubtful that such regulations will help achieve environmental goals in the long term. The increasing marginal costs of pollution abatement, the US industry's competitiveness problems, and the need to reactivate economic growth have made it urgent to improve the cost-effectiveness of the system. The trend is toward market incentives, with emission or effluent charges or pollution permits in some but not all areas.

All these changes will require institutional adjustments. Bureaucratic inertia may be one of the major obstacles in the way of greater efficacy and efficiency. Multimedia approaches will challenge the compartmentalized structure of the EPA. But the tension between decentralization and consistency and coherence — not to mention the loss of power — is another issue.

There is also a need to modify the public's perception. More and better information, improved communication channels, and new ways to involve the community are crucial to efforts to implement new policy concepts. A better understanding of the real risks and the trade-offs of environmental protection is fundamental to creating the necessary Congressional support. Participation will certainly be needed as the abatement efforts shift to small and nonpoint sources.

This section discusses some of the major changes affecting environmental policy-making, particularly in the case of mining-waste-disposal regulations. Recent proposals have introduced some of the new policy concepts, and it may be worthwhile for the reader to appreciate possible obstacles in the way of modernization.

Multimedia approach

At the beginning of the environmental era, the tendency was to react to the most pressing problems and the issues on the front pages of newspapers. Consequently, the approach was partial and had a media focus.

The main concerns in the early 1970s were air and water pollution — the most visible problems — so the first laws to be amended were the CAA, in 1970, and the *Clean Water Act* (CWA), in 1972. As the problems of air and water pollution were to some extent being resolved, new problems appeared, either because they had been overshadowed by previous emergencies, because new technological developments had created new problems, or simply because the media approach just shifted problems from one medium to another.

This partial vision had a strong influence on EPA's organizational structure, undermining its capacity to take a more integral perspective on environmental problems. The principal divisions of EPA were created following Congressional activity and reinforced the legislative pattern and fragmentary nature of US environmental policy (for example, see SAB 1990; Portney 1991). EPA's compartmentalized structure also created coordination problems, as well as inconsistency. Lastly, individual firms faced cumbersome and lengthy permitting processes.

These problems were already apparent in the early 1980s. The EPA (1984) noted that coordination problems often led to the duplication of research, inconsistent risk assessments for the same substance, the transfer of pollutants from one medium to another, and the uncoordinated regulation of the same industry by different programs. More recently, William Reilly, the new administrator of the EPA, stressed the shortcomings of the then current approach, particularly its negative impact on pollution prevention (Reilly 1989). EPA supported the Conservation Foundation's New Environmental Policy Project. The model developed by the Conservation Foundation requires the consideration of the environment as a whole in all decisions, a single-permit system, and the standardization of regulatory procedures (Irwin 1989).

Nevertheless, the change to a multimedia approach faced several obstacles, ranging from bureaucratic resistance to Congressional reluctance to award more authority and discretion to EPA. Industry was also concerned, fearing that changes in the status quo would bring new problems and necessitate new controls, with new costs.

Another consequence of a partial view was that it made it difficult to prioritize environmental problems and allocate resources to problems with higher health and environmental risks.

Risk assessment and prioritization

The objective of the Comparative Risk Project, developed in 1986, aimed to establish the risks currently posed by major environmental problems, given existing levels of control (EPA 1987). The study distinguished cancer and noncancer health risks and ecological and welfare effects and broke new ground.

The study's conclusions, based on the "informed judgement" of experts, were somewhat disturbing (EPA 1987). The ranking rather mismatched EPA's priorities, although the latter coincided with public opinion, reflecting the source of Congressional action. The best example of a discrepancy was the CERCLA program. According to the study, the risks associated with hazardous-waste disposal were rated very poorly. But Congress had reacted quickly and appropriated billions of dollars for the program's implementation.

The scientific basis for this first comprehensive attempt to assess the real risks was not as solid as one might have wanted. However, there was no doubt that the project was important, as the Science Advisory Board pointed out (SAB 1990). The report, in identifying the most significant risks, was an important step toward better allocation of limited resources.

EPA had used risk assessments on previous occasions when designing regulations. It had also used site-specific risk assessments when developing the National Priority List of CERCLA (Russell and Gruber 1987). But this was the first time that the concept had been used in a comprehensive way.

Risk assessment could be used to tailor standards and controls to local conditions and actual risks, improving the efficiency of the system (Tietenberg 1988). In addition, risk assessment provides a scientific foundation for identifying the social benefits of pollution abatement. One of EPA's goals is to impose requirements only where the benefits of regulation would outweigh the costs. However, it has proven difficult to design regulations that both meet this standard and are enforceable (EPA 1990). The second-best alternative is the use of cost-effective mechanisms, and this means an increasing reliance on market incentives.

Market incentives

The use of market incentives to internalize environmental costs of private decisions and reduce excessive pollution-abatement expenditures is getting increasing political support. The idea is not new. Economists have long been suggesting the use of taxes and marketable permits.

The Emissions Trading Program was the first attempt to introduce more flexibility into the ways environmental goals could be met. Concerned with the impact of future growth in nonattainment, Congress introduced this limited version of a marketable-permits system in the 1977 CAA amendments. The system awards emission-reduction credits to firms that reduce their level of emissions beyond those stipulated in the regulations. The firm can bank the credits and use them in the future for the same plant, or it can trade them to another company (Tietenberg 1988; Hahn 1989; Liroff 1989).

Nevertheless, not until the late 1980s did the concepts of market incentives and pollution permits find their way to Congress and the White House. The inclusion of a pollution-permit system in the 1990 CAA amendments was a landmark. Other market incentives have been proposed, including some to control CO_2 emissions to mitigate the greenhouse effect. Market incentives include the removal of barriers that prevent markets from working effectively and the elimination of government subsidies that stimulate the excessive use of natural resources.

Several economic and political factors explain why Congress and the White House endorsed the use of market incentives for environmental protection (Hahn and Stavins 1991). The economic recession of the early 1980s and the general slowdown of the economy had increased the marginal costs of pollution abatement. Easy targets had already been controlled; the next step would be to control the small and nonpoint pollution sources, which tend to present more complex problems. The technology for additional reductions of emissions and discharges implied higher abatement costs. Further economic growth also posed a challenge. Concerns about the international competitiveness of the US industry and the economy's capacity to absorb additional environmental costs motivated the search for more cost-effective mechanisms.

An important political factor influencing the use of market-based instruments during the Bush administration is that this kind of scheme fit well with the goals of the Republican administration. With the introduction of market incentives, the Bush administration was able to fulfil its commitment to environmental protection without intervening further in the economy and without imposing overwhelming costs on the industry or on the fiscal budget.

The introduction of market incentives was facilitated by the environmental movement's willingness to use economic tools in the search for better environmental quality. The Environmental Defense Fund, the Wilderness Society, and other well-known environmental groups have successfully used cost–benefit analyses to support their cause (see, for example, Stavins 1983, 1987; Goerold

1987). Their philosophy regarding the use of market incentives is pragmatic: if it works to protect the environment, let's do that. Nevertheless, the environmentalists are cautious, especially when economics is applied to more basic principles.

But the industry has not shown the interest one might have expected. It tends to be more conservative and to fear new rules for a game it already knows how to play. Administrative uncertainty and unexpected additional costs are at the root of that fear.

Although some analysts say it is premature to anticipate a massive use of market incentives, there is certainly a trend (Stavins 1991). In any case, there is consensus that the use of market incentives has to complement, not be a substitute for, the old system.

Institutional challenge

Institutional change is required if these concepts are to be incorporated into regulatory programs. The structure of EPA, its composition of human resources, and its budget need to reflect a higher degree of integration and a stronger role for economics. More administrative discretion may also be needed. This brings up a complementary subject: decentralization.

If a lack of flexibility stands in the way of more cost-effective ways of attaining environmental goals, a higher degree of decentralization will be needed. This will only work if state authorities have the commitment and resources to formulate their own programs and to monitor compliance. Theoretically, state authorities are in a better position to understand the specific problems and risks in their regions (at least, they are in a better position than Washington) and can tailor regulatory programs to the preferences of local communities and their willingness to pay for a cleaner environment. State authorities are also in a better position to monitor compliance and enforce regulations.

However, public-interest groups fear that giving greater discretionary powers to state authorities may mean a dirtier environment. The possible political alliances between local politicians and industry and the need to foster regional economic development may lead local authorities to soften regulations and standards, as well as enforcement.

Although the industry may benefit from this trend, it may also fear an excessively disparate regulatory system. This may be especially true of companies with plants in more than one state. Uniform regulatory programs are more expensive in terms of compliance but reduce administrative costs and uncertainty.

Finally, for certain problems, federal authorities cannot be replaced. These include interstate acid rain and water pollution and global problems. In some areas, important economies of scale and the need for a critical mass call for a stronger federal role, as in R&D activities or in the development of information systems.

RCRA amendments and mining-waste disposal

With the Bevill Amendment of 1980, mining wastes and certain mineral-processing wastes were temporarily exempted from RCRA regulations. The exemption was granted by Congress until EPA finished a study to determine whether these wastes should be classified as hazardous or nonhazardous. In 1985, EPA submitted the results of this study to Congress and, in 1986, published a regulatory determination on extraction and beneficiation wastes from mining. The principal conclusion of the study was that mining wastes should not be regulated as hazardous under Subtitle C of RCRA as originally proposed. Instead, EPA suggested a "tailored" approach for mining and beneficiation wastes under Subtitle D (nonhazardous wastes).

Since then EPA has worked intensely to produce draft regulations addressing concerns about the generation and regulation of mining wastes. The products, Strawman I and Strawman II, have been discussed by a variety of stakeholders: industry, environmental groups, the states, and federal agencies. The Policy Dialogue Committee (PDC), created in 1991, was the last chapter of this EPA effort. EPA intended to bring all interest groups to a public forum and eventually generate some agreements.

The reauthorization of the RCRA by Congress — which was to make a final decision on the legislative framework for handling these wastes and municipal and household wastes as well — was expected to take place in 1991/92. Although it was too early to predict the outcome of the EPA effort at the time of writing, some interesting aspects deserve attention: the unusual rule-making process itself; and the concepts in the regulatory proposals.

The most interesting feature of this process is the involvement of diverse interest groups in preparing and discussing the EPA draft regulations *before* Congress made its decision. Usually, public involvement takes place at three different stages in the legislative and regulatory process. First, the public has a chance to lobby and to bring expert witnesses once a statute has been introduced for Congressional discussion. Second, after Congress adopts a statute, EPA prepares and proposes the corresponding regulations to implement the statute; here the public may intervene by commenting on the proposed regulations. Third, after the regulations come into force, the public always has a chance to challenge in court the ways the law is implemented and enforced.

In the case of the RCRA amendments, though, instead of waiting for Congress, EPA took the initiative and started an informal rule-making process that had no precedent. This gave EPA the time and flexibility to involve the interest groups in the process. Moreover, the PDC gave these groups a chance to dialogue and interact. The groups may still consider each other as adversaries, but the communication flow — usually from each group to EPA — became multidirectional. This EPA initiative may have set an important precedent by making environmental policy-making in the United States less confrontational. This in turn might facilitate the implementation of regulations, reducing the legal and administrative costs and speeding up the whole process.

The concepts in the regulatory proposals included the use of a decentralized regulatory system, relying to an important extent on state-formulated programs; the development of programs on the basis of the real risks posed by mining wastes, instead of their potential risks; the use of site-specific controls; and the adoption of a multimedia approach, also a departure from the usual way of doing environmental policy in the United States.

Most of these concepts were present in the original formulation of RCRA regulations and are consistent with the new trends in environmental policy. Although the amendments contain no categorical statements, one may wonder whether they would have found their way into the regulatory language in the 1970s as easily as in the 1980s. The application of these concepts to mining-waste disposal has brought new light to them, particularly concerning the tensions and possible trade-offs of more cost-effective programs. Therefore, the experience may prove interesting.

In the 1990s, waste disposal will most probably be the main domestic issue on the US environmental agenda. For the mining industry, the reauthorization of RCRA and the approval of mining-waste regulations will be the next big regulatory step.

The problem

Around 4 or 5×10^9 t of waste is generated in the United States annually. An estimated 40% of this is from mining operations, including development, tailings, and leaching (Stone 1989). The other big waste generator is agriculture, contributing about 50% of the total.

Of the roughly 1×10^9 t of wastes produced by metals mining operations, 44% comes from the development stage; 33%, from tailings; and 23%, from leaching. Of the total, more than 50% comes from copper production (MacDonnell 1988). This is not surprising if we consider that more than 99% of the ore extracted is waste. Consequently, the copper-mining industry will probably be the most affected by the new regulations for mining-waste disposal.

Although these figures are impressive, the real risk posed by mining wastes is less thrilling. According to RCRA criteria, less than 25% of the $\sim250 \times 10^6$ t of hazardous wastes annually produced in the United States comes from mining and beneficiation. In the copper industry, most of the hazardous wastes (82%) come from copper-dump leaching operations.

Mining ranks second in the list of big generators of hazardous wastes. According to figures from the Congressional Budget Office (cited in Dower 1991), 48% of US hazardous wastes are generated by the production of chemical and allied products, and 18% are produced in the primary metals industry. Petroleum and coal products generate another 12%.

Although mining wastes pose some degree of environmental risk, particularly to groundwater, they differ from other industrial hazardous wastes, as well as from municipal and household nonhazardous mining wastes. Mining wastes

- Come in higher volumes, especially compared with the volume of the associated products;

- Cover large area;

- Are disposed of at the site where they are generated, thus involving no transportation of hazardous substances;

- Are usually disposed of in dry and sparsely populated areas;

- Consist, to an important extent, of unprocessed waste; and

- Pose lower risks.

History of RCRA and mining-waste-disposal regulations

RCRA was formed in 1976 in response to public alarm over hazardous-waste sites. RCRA combined two previously existing regulations, the *Solid Waste Disposal Act* and the *Resource Recovery Act*, and was intended to provide the EPA with the authority to regulate, control, and monitor hazardous substances. Two years later, in 1978, EPA proposed rules for hazardous-waste management under Subtitle C, creating a special-waste category to include mineral-industry wastes. EPA's intention was to give some flexibility to the industry in the treatment of these wastes, given their special nature (Kimball and Moellenberg 1990).

Nevertheless, in the regulatory document that EPA submitted to Congress in 1980, mining wastes were practically subject to the same requirements as those affecting industrial hazardous wastes, with the exceptions only of overburden used for reclamation purposes and *in situ* mining wastes. EPA also proposed to list several mining-processing wastes to be regulated as hazardous wastes under the same Subtitle C.

Congress, aware of the particular characteristics of mining wastes and concerned about imposing unnecessary costs on the industry, prohibited EPA from applying hazardous-waste regulations to solid wastes from extraction, beneficiation, and processing of ores and minerals until the completion of the detailed studies of these wastes. This was the so-called Bevill Amendment, introduced in the *Solid Waste Disposal Act* amendments of 1980. The deadline for the studies was 1983. As a consequence, except for those hazardous wastes not deemed unique to the mining industry (that is, chemical substances), mining and processing wastes were temporarily exempted from RCRA regulations. At most they were subject to state regulations.

In 1984 several environmental groups sued EPA for failing to meet the deadline (*Concerned Citizens of Adamstown* v. *EPA*). They also challenged the inclusion of mineral-processing wastes in the Bevill Amendment. In response, EPA scheduled the completion of the studies and limited the number of mineral-processing wastes to be exempted.

In December 1985, EPA submitted the study to Congress. The report concluded that regulation of mining and beneficiation wastes under Subtitle C of RCRA was unwarranted. However, acknowledging some potential risks, EPA decided to develop a program under Subtitle D (nonhazardous wastes).

Given the original objective of Subtitle D — to regulate municipal- and household-wastes disposal under state supervision — EPA suggested a special program tailored to mining wastes. EPA was concerned about the need to take into account the fact that the risk varied from site to site, depending on the characteristics of the particular mining wastes and on local environmental factors, such as climate, geology, hydrology, and soil chemistry. Consequently, EPA proposed a flexible, site-specific, risk-based program (Housman and Walline 1990).

Another EPA concern was that the responsibility for administering Subtitle D of RCRA had been left to the states. EPA suggested a stronger role for federal authorities to ensure human health and environmental protection.

The 1986 report failed to address the issue of mineral-processing wastes from either abandoned or inactive mine sites. A decision was made in May 1991 regarding the 20 mineral-processing wastes subject to the exclusion, following a 1988 court order that restricted the interpretation. Eighteen were kept under Subtitle D. The other two were made subject to Subtitle C, CERCLA, and the *Toxic*

Substances Control Act. The mineral-processing wastes removed from the exemption and those never listed are subject to Subtitle C if they have hazardous-waste characteristics. If not, they may be regulated under the new program under Subtitle D.

Since 1986 EPA has been working to develop a regulatory program through its Office of Solid Waste (OSW) and Region VIII (its regional counterpart). The same year as EPA released its report, it established a Mining Waste Regulatory Development Workgroup, with members representing EPA offices and federal agencies. This workgroup acted as an advisory group for the OSW.

In 1987, EPA established an External Communications Committee, consisting again of representatives from EPA and other federal agencies. Its role was to foster communication among all interested parties, including state agencies, industry, and public-interest groups.

In 1988, EPA released Strawman I, a set of draft regulations developed jointly by the OSW and Region VIII. Understood to be a working paper, Strawman I was to serve as a starting point for discussion. After receiving written and oral comments from the interest groups, the OSW and Region VIII prepared a second version, Strawman II, published in 1990.

In 1991, EPA officially created the PDC to bring all interested parties together to exchange points of view. Each group — whether state, federal, industrial, or environmental — has seven representatives on the PDC. The Keystone Center has acted as an independent facilitator for the meetings. The PDC meets every 6 weeks, and the meetings and their minutes are open to the public.

Group involvement and the PDC

Environmental policy-making in the United States has been extremely confrontational. In part, this is a consequence of US political culture and the common use of the judiciary system to solve disputes. Thus, EPA's efforts to involve interested parties from the very beginning and to reach some degree of consensus are especially interesting.

Usually, EPA prepares draft regulations after Congress enacts a piece of legislation. In this case, the OSW suspected that the reauthorization of RCRA — which would have triggered the normal process — would take some years. So, the OSW took the initiative to obtain inputs from all the interested parties right from the beginning. The OSW's objective was to create a regulatory program that all parties could live with.

The OSW had another purpose in mind. Under conventional circumstances, EPA cannot influence Congressional decisions, as other players might, by lobbying. Congressional and White House approval of a bill is a very political process.

The creation of the PDC gave EPA an alternative for influencing decision-making at the approval stage. The spotlight on the issues and the interaction with the other interest groups gave EPA better access to the political scene.

To avoid bureaucratic deadlock, EPA opted for an informal process that would not require the endorsement of high management levels. The first step was Strawman I, a first draft prepared by the OSW and Region VIII in 6 weeks. The purpose of the document was not to deliver EPA's final word on mining wastes but to stimulate discussion.

As part of the effort to encourage the public to participate, EPA gave financial support to a number of groups in 1988 to analyze the problem and respond to Strawman I. With this funding, the Western Governors Association (WGA) formed a mining-waste task force; 21 states participated (Housman and Walline 1990). Also participating was Colorado Trout Unlimited, formed in 1990 by several prominent environmental groups, such as the Environmental Defense Fund and the Mineral Policy Center. Finally, EPA also supported an association of small-scale miners, the Northwestern Mining Association. The American Mining Congress represented medium- and large-scale mining companies in the discussion.

This support — as well as the focus of Strawman I on practical issues, rather than on regulatory principles — stimulated and facilitated the participation of the groups. Thanks to this approach, EPA received input from the industry, environmental groups, federal agencies (such as the USBM), and the states (under the umbrella of WGA). Public hearings were held, as well as informal meetings, and USBM, WGA, and the American Mining Congress prepared written comments. Informal channels between EPA and other parties were also used.

After 2 years of discussion and work, EPA published Strawman II, a review of Strawman I that incorporated oral and written comments received. It was closer to a final draft, but EPA still invited discussion. As usual, environmental groups charged EPA with being too lenient, and the industry complained about the rigidity of Strawman II — according to the industry, it was closer than Strawman I to Subtitle C. In this second round of discussions, the OSW realized that an important part of the problem was the lack of understanding each group had of the other parties' concerns and that each one's strategy was basically to recover lost ground.

To overcome this impasse, the OSW proposed to the EPA Deputy Administrator that the PDC be set up under the terms of the *Federal Advisory Committee Act* (FACA). The purpose of FACA is to sanction external advice given to governmental agencies and to prevent unnoticed outside influence. FACA had been used before, most often to form regulatory-negotiation committees. These committees had to be established by statute, and the participants had to agree to waive their right to sue each other over the agreements achieved. In this case, there was no statute and the OSW preferred to avoid pressuring the groups to make final agreements. The alternative was to form the PDC, which would have none of these requirements. The Deputy Administrator of EPA approved the OSW proposal and the PDC constitution in April 1991 (EPA 1991).

The PDC gives a great degree of freedom to both the OSW and the groups. The agreements of the PDC do not require the support of EPA's Administrator, and the groups are not forced to reach agreements that compromise their future actions. With the PDC, direct compromise and some consensus are possible. An agreement is a powerful signal to Congress, although such an agreement has no resolutory status.

It is too early to comment on the success of the PDC, but all the groups agreed that participating on the PDC improved their understanding of each other's concerns. Certainly, this was one of EPA's main objectives in setting up the group. The PDC has also focused the debate on specific issues and provided an equal standing to the different parties in the discussion.

The groups seemed less optimistic when asked about possible agreements. At the time of writing it looked like the positions of state and federal agencies and the industry were getting closer, whereas the environmental groups were lagging behind. Several factors may explain, in part, the difficulties in reaching some consensus. For one thing, unlike members of a regulatory-negotiation committee, these players had no real authority to make decisions. If the PDC members are able to reach some agreements, Congress may take these into consideration but only as advice. This feature theoretically increases the freedom of the players, but it diminishes their confidence in the PDC's ability to influence policy decisions. I say "theoretically increases the freedom of the players," because the use of a public forum puts different groups in the spotlight — especially environmental and industry groups — and forces them to emphasize principles over concrete issues. Extreme positions tend to be favoured over pragmatic compromises because of the fear of diluting the message in an attempt to find intermediate positions.

Some participants expressed their apprehension about the Keystone Center's role as facilitator. They did not consider the Keystone Center a truly impartial facilitator, as it has a contract with EPA and has to follow its guidelines.

The size of the group does not facilitate interaction among the different parties. This obstacle may be overcome to some extent through the recent creation of subcommittees to discuss specific issues.

Informal pre-meetings of some of the PDC members have reinforced the natural distrust between the groups — the groups that did not attend fear the development of covert agreements.

The problem is too broad, involving too many fundamental issues. Too much is at stake for each group. Although the advisory character of the PDC diminishes the pressure to some extent, the groups still feel that agreements will entail important public compromises and set possible precedents.

The PDC has probably brought the two extremes of environmental controversy to the table. The gap between the industry and the environmental groups appears very wide, which is partly a consequence of powerful images built up in the past. The mining industry looks at environmental groups as if they were concerned only with birds and bunnies, and environmental groups consider mining the most backward industry in terms of environmental responsibility. Each group recognizes that there is a broad spectrum on each side, and it is not clear what the position of the representatives in that spectrum is.

The reauthorization of RCRA by Congress has stimulated parallel lobbying. The different groups are aware of this phenomenon, reinforcing their lack of confidence in the PDC's ability to produce concrete outcomes.

Finally, the PDC was established after 3 years of discussion of the issues, and some participants feel frustrated because the PDC is bringing the discussion to the starting point again.

Some of the obstacles mentioned by the interviewees may be overcome in the future with a different design for the PDC and its meetings or with better timing. Other obstacles may require more substantial efforts, as distrust appears to be an important component. However, this kind of initiative may be extremely useful, regardless of whether it achieves more tangible outcomes.

The issues under discussion

Several issues were under discussion in the Strawman I and Strawman II periods, as well as during PDC meetings. In the following pages, I examine the most controversial issues, particularly those related to the new trends in environmental policy-making. They are discussed separately, although all are strongly related.

STATE VERSUS FEDERAL AUTHORITY — The distribution of power between federal and state agencies is probably one of the main issues. To an important extent, this is the institutional counterpart of uniform versus site-specific regulations. The state authorities are better prepared to evaluate the specific environmental impact of a mining site. They have first-hand knowledge of the conditions in which the companies work — both the operational characteristics and the environmental setting.

Many states already have programs to ensure adequate management of mining and minerals-processing solid wastes. These programs cover aspects like groundwater, land reclamation, dam safety, and financial assurance (Housman and Walline 1990). These programs vary from state to state, according to variations in climate, geology, and environmental sensitivity of the impacted areas (Stone 1989). It seems inefficient to disrupt existing programs.

Various EPA documents have recognized the important role that the states play in formulating and enforcing specific regulations. EPA is interested in maintaining its flexibility. It wants its programs to be compatible with state programs, and it wants to give the states a leading role in developing, overseeing, and enforcing their own mining-waste-management plans (Housman 1990).

EPA is still responsible for protecting human health and the environment. If a state has not developed a special program or does not meet minimum federal criteria, EPA needs the authority to go beyond the general guidance and assistance guaranteed under Subtitle D.

The industry, USBM, and WGA support a stronger role for the states in the design and implementation of programs. The industry and USBM believe that state agencies are better acquainted with mining specificities. Mining activities are highly concentrated in a few states, and those states have ample experience dealing with the mining industry. The industry also wants to see a better delineation of authority to avoid "having to serve two masters." States want to maintain their current programs and their authority.

Environmental groups fear that too much discretion on the part of the states will result in insufficient environmental protection. They want to see a stronger role for EPA. Their arguments are diverse. They fear that state authorities may be influenced by industry to set softer standards and loosen monitoring and enforcement programs, especially in states where mining is an important source of income. Although a state may be genuinely committed to an environmental program, it may not have the necessary resources and capacity to establish and enforce this kind of program.

EPA has proposed that plans be approved by EPA. Once the plans are approved, EPA's regional offices would have oversight and enforcement authority in the states, and EPA would issue and enforce permits in nonapproved states. EPA also suggested that it would intervene whenever human health or the environment is at especially high risk.

The conflict between EPA and the supporters of greater state discretionary power is really about some ambiguities in EPA proposals, especially in those paragraphs giving EPA authority over state programs in special circumstances. "EPA could usurp regulatory authority from the states at any time" (Kimball and Moellenberg 1990; USDI and USDA 1990).

FLEXIBILITY AND UNIFORMITY — Tailoring a program to site specifics is an important departure from traditional regulatory programs. EPA recognizes that the benefits of environmental protection (or the damages of no protection) depend on the specific environmental conditions. Instead of trying to apply an across-the-board regulatory program, EPA is attempting to design a program on the basis of the real risk posed by mining wastes, getting closer to the ideal scheme. EPA is also putting more emphasis on balancing those benefits with the costs imposed on the industry.

These concepts were in EPA's draft regulations and helped EPA gain Congressional approval for the Bevill Amendment. Although EPA's language insists on the idea of flexibility, EPA bases its proposed standards on ongoing regulatory programs and considers design and operating criteria that run counter to the original spirit of flexibility.

EPA's groundwater standards are designed to match the maximum contaminant levels of state programs established under the *Safe Drinking Water Act*. If these data are unavailable, a health-based risk-assessment standard is used. If neither of these standards is available, background levels become the criteria.

The industry considers the performance standards of Strawman II to be inflexible and even more stringent than those of Strawman I and closer to those of Subtitle C. USBM and the industry want to restrict EPA's role to that of providing technical guidelines — they do not want EPA to impose specific technologies.

On the opposite side, environmental groups consider flexibility risky; they want national minimum-performance standards. They argue that flexibility and state discretion imperil environmental protection because it may become a source of competitive advantage for the states. A flexible program is also more difficult to monitor and enforce because it requires higher administrative capacities and more resources. The environmentalists want to see a prescriptive and detailed program.

This partly explains why environmentalists want a statute provision allowing citizens to sue companies with unsound environmental practices. Without this provision, severe environmental damage has to occur before citizens can sue a company.

MULTIMEDIA APPROACH — The multimedia approach is another shift from the traditional approach. Although mining is already subject to several environmental regulations, including those of state programs that deal with mining-waste disposal, EPA is concerned about some remaining gaps (Housman 1990).

EPA wants RCRA amendments to cover the whole spectrum of mining-related environmental problems. A multimedia approach would also have administrative advantages: for example it would avoid duplication and conflicts between different regulations, reduce the administrative burden, identify possible disincentives, and give regulators a better picture of what is going on. EPA proposed the idea of a one-permit system incorporating all current regulations plus new ones. The states are expected to design a multimedia approach that addresses air, water, and soil contamination and incorporates existing permit requirements such as those required by the CAA (Housman 1990).

EPA wants to extend the scope of the regulatory program to include exploration wastes and materials that are not necessarily waste, such as those related to heap-leaching operations and abandoned mines. If active leaching piles — considered operating units, not waste — are left unregulated, companies might extend the life of these piles simply to avoid the cost of regulatory closure. The exclusion of abandoned mines from this program might provide a disincentive for recycling and, in general, the disincentive for a more effective cleanup of hazardous sites (Peterson n.d.). EPA intended to include incentives to mine mining wastes wherever there may be a net gain for both the environment and the company. EPA is aware that because of the present structure of CERCLA, the development of incentives will require streamlining of the regulatory process and revision of current operating and performance standards (Housman 1990). Environmental groups welcome the multimedia approach because it reduces the possibilities of gaps and of shifting the problem from one area to other.

However, the industry definitely opposes a multimedia approach. Although it makes sense for the industry to resist new standards and controls, the industry also opposes a one-permit system, a system that would reduce administrative costs. The industry argues that changing the system will introduce uncertainty and that the net costs of the change are unclear. The industry prefers keeping a system that, if not perfect, is better known and will bring no additional surprises.

USBM and the states are in a mixed position. USBM wants a single regulatory program for all wastes, including processing wastes, to reduce the administrative burden and increase consistency. USBM also emphasizes that re-mining of old mining-waste sites and impoundments and recycling of materials to reduce hazardous waste should be encouraged. But USBM opposes the one-permit system and the inclusion of exploration wastes and heap-leaching materials. USBM maintains that if there are gaps, it is because current laws are ineffectively enforced, and it fears the duplication of authority and programs (USDI 1990; USDI and USDA 1990). The states agree with the inclusion of heap leaching in the program but are more cautious about the multimedia concept. They do not want to disrupt current programs and think that having a comprehensive permit will affect current institutional structure.

In Strawman II, EPA gave up on the idea of a one-permit system and left the issue of abandoned mines for future amendment of CERCLA. Heap leaching is still under discussion.

CRITERIA FOR STANDARDS — Many issues related to criteria for standards have generated discussion. The most important of these issues are compatibility with existing standards, like those of CAA or CWA; the distinction between old and new facilities; and the use of technology versus performance standards.

As expected, the industry opposes stricter standards and, whenever possible, prefers to keep the primacy of current standards — especially the more lenient ones, such as those for groundwater quality — and the use of background levels. Inertia also explains some of the resistance to new standards. Although the industry struggles to avoid additional or more stringent standards, the position of environmental groups is that there should be no degradation. The industry is also concerned that compliance with new standards may not be feasible or economical in the case of old facilities. The states have proposed a deadline for old facilities to comply. Finally, the industry and USBM prefer having performance standards, instead of technology standards, because of their effects on innovation. USBM emphasizes that the industry should be allowed to find less expensive solutions.

Final comments

It remains to be seen whether the EPA initiative provided useful inputs to Congress and helped to shape more meaningful and realistic statutes. But it is at least useful to reveal some of the tensions and obstacles in the way of more flexible and cost-effective programs and consensus.

We may see that to an important degree, the trade-off between more flexible programs and effective environmental protection rests on the real independence of local authorities and their capacity to implement and enforce the regulations. Enforcement failures and consequent environmental degradation may outweigh potential benefits. The political structure of the United States enhances the struggle for control between the regions and Washington. This tension is heightened by the extreme positions adopted by the industry and the environmental groups. Both have taken RCRA amendments as their trench to defend their dearest positions, increasing the usual distance. The industry strategy is to bring all new regulations under the RCRA umbrella to avoid new regulatory initiatives in other battlefields. The approval of RCRA amendments is the industry's opportunity to protect itself from new laws. Environmental groups see RCRA as their best opportunity to get a comprehensive regulatory scheme for the mining industry. Other initiatives, like modification of the *Mining Law*, will probably take longer to be approved because the issues are more fundamental and controversial.

Another big obstacle is industry and bureaucratic inertia. The discussions have made industry inertia especially patent, confirming the generally conservative attitude of the mining industry and its aversion to taking risks and trying new approaches. The industry is particularly sensitive to uncertainty. Bureaucratic inertia will probably become apparent after the implementation of a program.

Finally, ambiguity is an important barrier to agreement. The different sides tend to interpret procedural or substantive ambiguities to support their prejudices or fears. This tendency hardens positions and make things more difficult than they really need to be. The Strawman and PDC exercise helped to expose this problem.

Policy recommendations

Certainly, environmental regulations have had an effect on the US mining industry's profitability. Companies have been forced to retrofit or renovate installations or leave the market. Increasing operational costs have affected their international competitiveness, and to some extent, this may be changing the world allocation of mining investment. Employment levels have fallen substantially, and local economies have borne part of this cost.

Environmental regulations have also brought with them important benefits: better air and water quality and reduced health and environmental risks. It is unclear whether the benefits compensate for the costs, much less whether the net social benefit is maximized. The aggregate numbers may show positive net results, but at a regional or local level the situation may be very different. Sometimes it is clear that the same results could have been achieved by spending less.

This does not necessarily imply a negative assessment of efforts so far or a denial of some important achievements. I wish to emphasize the trade-offs in environmental policy-making and the difficulties in measuring these trade-offs — gains against costs — and in making rationally optimal decisions about the appropriate levels of pollution control and environmental protection.

By learning lessons from the US experience, we can make our own process in Chile less painful, more efficient, and more effective as we strive to improve both the quality of our environment and our chances of keeping on a sustainable-development path.

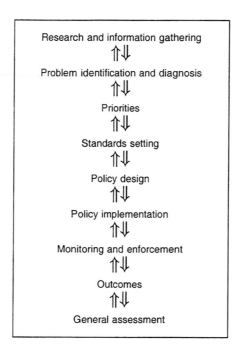

Figure 2. The policy-making process.

If one were to draw a flow chart to illustrate the policy-making process, it might look like Figure 2. The arrows emphasize the dynamic and interactive nature of a process that is never really complete. All the steps shown, particularly research and information gathering, are essential to achieving environmental goals, but here I concentrate on policy design and implementation (although, actually, policy design encompasses design at all stages). I first discuss the main criteria for assessing the desirability of a policy and some concepts that should be at the basis of policy design. I will then discuss tactical issues like instrument choice and institutional aspects, keeping my analysis to a general level, although I will discuss the specifics of mining regulations when relevant.

Main criteria for policy assessment

Three main criteria can be used to assess the desirability of a policy:

- *Effectiveness* — Are we going to be able to reach the goals we have set? This is an obvious criterion, but experience shows that it is not always easy to meet. The reality is complex and evolves over time. Are we introducing dynamic considerations, such as economic growth and its impact on pollution levels? Are our goals too ambitious, leading to the inevitable extension of deadlines and the eventual abandonment of original programs, undermining our credibility? Are we overlooking relevant second-order factors, like indirect disincentives to recycle?

• *Efficiency and cost-effectiveness* — How close will policy outcomes be to the socially optimal levels of pollution control? Are we to the greatest possible extent taking into account the benefits and costs of environmental controls? Given a specific standard, is this the least expensive way to achieve it? Are we taking into account compliance costs, control and monitoring costs, enforcement costs, and general administrative expenses?

• *Political feasibility* — Will a policy have enough political support? How long might it take to get approval from the legislative and the executive? How much resistance will there be in the implementation stage? How fair will the various interest groups perceive the policy as being? What are the chances of long-term stability?

From an examination of US experience, five central concepts emerge in the design and implementation of an effective, efficient, and politically feasible environmental policy.

An integral, multimedia, ecosystemic perspective

An ecosystem is a set of interdependent organisms in an ongoing process of adaptation to their environment. An ecosystem is described in terms of its biological elements, their mutual relationships, and their relationships with the physical and chemical media that support life.

When we are concerned about environmental quality, we are concerned about the disruption of these relationships and the endangerment of the capacity of the system to adapt, evolve, and survive. So it seems logical to use a systems approach to confront the problem. This is the only way to take into account multiple relationships, impacts of several orders, and synergistic effects. It is the only way to be really effective.

We have already seen that taking a partial view of environmental problems creates new and sometimes worse problems. Other negative practical consequences are difficulties in prioritizing problems, resulting in inefficient allocation of available resources; duplication of effort; missed opportunities for economies of scale and positive indirect impacts; and a complex and cumbersome bureaucratic structure.

We may call this multimedia or systems perspective an integral perspective, as opposed to a partial one. It is fundamental to have this integral perspective from the beginning, instead of thinking of superposing different programs at a later date. The perspective taken at the start will not only affect the perception of opportunities to attack the problem but also avoid the creation of inertial forces and interests among the different actors in the system, as happened in the United States.

Flexibility

Policymakers, industry, and the public have to be aware that conditions will be changing and that the regulatory system has to be flexible enough to adapt to new circumstances. Additional knowledge and information, new technologies, and new socioeconomic conditions will necessitate modifications to priorities and strategies. Moreover, the environment is constantly evolving, and the problems will change. The ongoing assessment of policy outcomes should provide a feedback process to ensure that the regulatory system adapts properly to changing circumstances and remains effective and efficient (see Figure 2).

The regulatory system should be flexible in yet another sense. Once the system is established, the main concern is in meeting ambient-quality standards (these standards should be emphasized more than emissions standards). Whenever possible, firms and individuals should be given discretion to choose how best to meet these standards. This is the best way to ensure minimum compliance costs and also to promote technology development.

However, the need for flexibility has to be balanced against the need of economic agents to reduce uncertainty. This is a particularly strong concern in the mining industry. Clear rules and appropriate phasing of programs will be central to balancing these needs.

Specificity

Environmental problems differ from region to region. The types of pollutants emitted or discharged and the capacity of the environment to cleanse itself vary. Population density and the degree of exposure also vary geographically. Negative externalities depend on the kind of economic activity pursued. Environmental impacts are specific to an ecosystem and its demographic and economic conditions. This is especially true of the environmental impacts of mining activities.

On the other hand, people's preferences for environmental quality and other goods depend on socioeconomic and cultural factors that are equally diverse.

The more we are able to tailor environmental programs to the actual problems, the more effective we will be in reaching our goals and the less we will have to spend in doing so.

This is consistent with a multimedia approach. As the complexity of the system increases and we try to go from thinking of the parts to thinking of the whole, we may want a different way of simplifying the real world: reducing the geographical areas to which certain parameters apply. The limits of a properly defined ecosystem may be the alternative we are seeking.

Participation

The public has the last word on the importance and adequacy of environmental policies. Communities should play a substantive role in the policy-making process to ensure equity, effectiveness, and efficiency of the regulatory system. The community's preferences for a cleaner environment and ecological preservation are the ultimate criteria that shape the benefits of environmental protection.

On the other hand, to increase the feasibility, effectiveness, and efficiency of a policy, diverging interests should be reconciled by avoiding confrontational dynamics and using negotiation. Confrontational dynamics are time and resource consuming. Involving all the interested parties ensures fairness; legislative expedience; reduced resistance and litigation in the implementation phase; and increased likelihood of long-term stability.

A public that has been manipulated by the misuse of information is not well prepared to assess environmental problems or to judge policy matters. Lack of awareness about health hazards and ecological risks on the one hand and biased and inflamed discourses in favour of environmental protection on the other distort public opinion and diminish people's capacity to make a proper assessment of alternatives. This is why objective information and public awareness are so important.

The public should also be aware of the costs of environmental protection and of the consequences of their everyday actions. Few people realize that environmental problems and solutions are tied to daily decisions and that costs are going to be borne by the whole of society. A community must be well informed so that it can make responsible decisions about how much economic growth it may have to sacrifice in order to enjoy better environmental quality.

Pragmatism

The feasibility and effectiveness of a policy depend on how realistically it is framed. Strict programs that go beyond the real capacity of the industry to comply result in extensions that undermine the credibility of the new deadlines and the technical capacity of the agency and implicitly justify leniency in enforcement of legislation.

Ambient-quality standards should be defined on the basis of human health first and welfare and ecological considerations later on. However, specific programs for reaching those standards, phases, and deadlines should be based on considerations of economic and technical feasibility, with market considerations included. Timing is key to avoiding unnecessary costs and to reducing industry resistance.

Instrument choice

We are looking for a flexible, integral, participatory policy to insure effectiveness, efficiency, and feasibility. What are the instruments that best serve our purposes?

Before there were environmental policies, the courts were the only recourse for those affected by environmental problems. This proved to be ineffective and inefficient because of the uncertainty surrounding court decisions. The law left too much scope for interpretation. It was also expensive, and transaction costs were prohibitive for some of those affected. The public-good nature of this solution generated the problem of free riders. Finally, information was scarce, and secret settlements precluded the diffusion of information relevant to other actors. Of course, the use of courts to address environmental problems depends on a well-developed and accessible judiciary system.

As we have seen, although command-and-control regulations significantly advance the cause of environmental protection, they have been open to many criticisms. Some regulations have been attacked more than others (for instance, design versus performance standards), and the system has been accused of being rigid, bureaucratic, and expensive.

The same voices of criticism have advocated both the use of economic incentives (taxes and marketable pollution permits) and, in a more general sense, the use of the market to correct failures and eliminate distortions. To be sure, with market incentives one still needs regulations, but the mechanisms used to ensure that standards are met are different.

There is another alternative — the Coasian solution. This means leaving the interested private parties to negotiate over the problem, with property rights well defined. This theoretically optimal solution has been mostly kept within the pages of textbooks because its appeal depends on rather unrealistic assumptions. Normally, the negotiated outcome differs from the optimal solution because of transaction costs, strategic behaviour, manipulation of information, and income effects. Also, although theoretically efficient, this approach fails to address equity issues, like intergenerational concerns and the distribution of property rights. Uncertainty is another shortcoming of this approach.

Emphasis should be given to the role of ambient-quality standards in controlling the environmental impact of economic activities. It is preferable to set general targets for pollution abatement and to allow the allocation of the targets among the pollution sources to be driven by economic instruments such as taxes or permit prices. If this is impossible, then setting performance standards is still preferable to setting design standards. Also, where relevant, intermittent controls, depending on, for instance, meteorological conditions, are preferable to permanent controls. The mechanisms that we choose to ensure that our environmental-quality targets are met should at the same time allow for economic growth and be able to adjust to new conditions smoothly. Economic instruments (like taxes or permit auctions) may be preferable because they raise funds to support the whole system. These preferences have to be balanced against considerations of technical feasibility, such as the capacity to monitor compliance and enforce regulations and the ability to meet environmental-quality targets without creating hot spots or long-range problems like acid rain.

Looking at the situation in the United States, one might be tempted to jump into marketable pollution permits to achieve environmental standards:

- They ensure that environmental goals are met;

- They give firms the flexibility to choose their own strategy to comply while minimizing their costs;

- They adjust to economic growth and inflation; and

- They provide important dynamic incentives for developing new environmental technologies.

Marketable pollution permits seem just perfect; nevertheless, we have already recognized some caveats. The most obvious has to do with the case of substances that are toxic and pose big health and environmental risks at even low concentrations. In such cases, we cannot allow firms to decide the appropriate level of pollution abatement on the basis of their particular costs.

With this exception, three main considerations determine whether it is worthwhile to use marketable pollution permits. First, although the use of pollution permits instead of command-and-control regulations may increase net social benefits, the gain greatly depends on the competitiveness of markets. The gain is not as great in highly concentrated markets (either the permits market or the product market) as in atomized markets. Still, if pollution permits are used in markets in which they become a serious entry barrier, they may result in losses of efficiency that counteract the advantages of the system.

Second, the costs of environmental protection include not only the compliance costs of the firms but also the control, monitoring, and enforcement costs and the general administrative costs of setting up the system in the first place. The use of pollution permits (particularly ambient-quality permits, as opposed to emissions permits) is theoretically the most efficient approach but requires advanced modeling techniques and an especially good system of monitoring and compliance. These requirements limit the economic and technical feasibility of implementing a system of pollution permits.

Finally, marketable pollution permits will not work as well as command-and-control regulations if the pollution problem is so acute that it requires maximum control.

When choosing specific instruments, it is crucial that we look at the characteristics of the environmental problem; the specifics of the sources of the problem; and the social, economic, and political conditions we face. In this sense, the instruments we choose have to be appropriate for specific situations, and most likely, a mix of different instruments will be needed to reach global goals.

Having said this, I wish to call attention to the use of case-by-case negotiations for controlling the impacts of medium- and large-scale mines — the big point sources — and propose a modified version of the Coasian solution. The negotiations will take place after a general framework has been established. The framework will comprise delimitations of ecoregions; ambient-quality standards, related global abatement targets, and maximum emissions ceilings; clear definitions of property rights; decentralized authority and resources; empowerment of local communities; and clear rules for negotiation. Environmental-protection specifics (such as how the standards will be met) and specific programs (with specific degrees and types of control and timing) will be left to negotiations between the industry, local community, and local authorities. Ideally, the negotiations will cover all the relevant issues at once, giving rise to a single permitting process. The outcomes will reflect community preferences, specific environmental conditions, and the particular conditions of the firms and will address equity, effectiveness, and efficiency concerns. Of course, this scheme will be valid for either local or regional environmental impacts.

Central authorities will have the main responsibility for defining the framework and will generally oversee the process, including monitoring of compliance and enforcement of ambient-quality standards and negotiated agreements. Equally important, the environmental agency will have a central role supporting the development, analysis, and diffusion of information deemed relevant to the negotiations. Finally, I emphasize the importance of having the appropriate financial and human resources to undertake the task. These will include properly trained government officials and community and industry representatives.

Environmental problems generated by small mines, mainly groundwater and surface-water pollution, will require different schemes. Also, global problems will require a national strategy.

Institutional aspects

An integral, flexible, and participatory approach requires a great deal of multidisciplinary work, as well as interaction and coordination of various government agencies and private institutions.

Each and every ministry and government agency must become sensitized to the environmental impacts of economic activities and must consider the environmental impacts of every policy and program. The use of EIAs has become a normal practice in the United States and has had an important impact. However, it is necessary to go beyond bureaucratic mechanisms that may become just part of a red-tape ritual.

An approach like the one I have just proposed also requires a great degree of decentralization. Although central administration is required to secure at least a minimum degree of environmental quality and uniformity, most of the decisions can be made at the regional or local level, depending on the scope of the environmental problem. This will require the empowerment of regional and local authorities and the promotion of community organizations. The delegation of authority has to be matched with the appropriate human and financial resources to ensure the capacity of the local agencies to deal with the problems.

The Ministry of Mines

The institution ultimately responsible for designing, implementing, and enforcing environmental policies should be a special environmental agency with the independence and expertise necessary to undertake the effort. The Ministry of Mines has nevertheless a very important role in representing the interests of the mining sector according to long-term mining-development policies. The Ministry of Mines will be the main contact between the environmental agency and the mining sector and will be the principal source of information. The ministry will also be the main source of support for the industry, to minimize compliance costs.

The Ministry of Mines should develop appropriate channels of communication with the environmental agency to ensure that policy decisions are well informed and take into account factors affecting the development of the mining sector. Collaboration and policy coordination play a very important role in the effective use of available resources.

Another important area of work is the development of information and policy analysis. The assessment of the actual environmental impacts of mining, as well as of the impacts of environmental regulations on the mining industry, is a basic step and fundamental input for policy-making. Such assessments will require the establishment of channels of communication with the industry, the installation of monitoring equipment, and the development of modeling techniques. To evaluate the capacity of various segments of the industry to comply, international-market conditions and technology availability should be considered. Imagining innovative ways to fulfil environmental goals and formulating strategies to compensate for potential losses of competitiveness and employment are some of the challenges facing the ministry.

The development of new environmental technologies should be a central element in the promotion of R&D. The future competitiveness of the industry may depend to an important extent on its capacity to develop less expensive pollution-abatement technologies. For instance, SO_2-fixation techniques, alternative uses for H_2SO_4, and hydrometallurgical techniques should have an important place on the agenda. The Ministry of Mines can also promote technology transfer, especially in the case of small-scale mining. Extremely important is the development of training programs to create the required expertise in the different areas and at the different levels. Finally, the Ministry of Mines can improve the linkage between the domestic mining sector and international expertise, promoting exchange programs with and technical assistance from those countries with more experience.

Final comments

It would be inconsistent with the principles expressed here to go much further in making specific policy recommendations. These depend on the diagnosis and the particular conditions where the problems are experienced. For the copper industry, all I can say is that the most expensive environmental regulations will be those governing SO_2 emissions and mining wastes. Consequently, special attention should be given to assessing the possible economic impacts of these regulations and to fostering the development of cost-effective alternatives for dealing with these problems.

A scheme like the one I proposed will be feasible in a country like Chile: the administrative division of the country may be useful in establishing an eco-regional approach, and the small number of big mining firms will make it easy to use a case-by-case approach and to enforce negotiated agreements. It seems evident that Chuquicamata should not be subject to the same degree of emissions control as Ventanas or Chagres.

Chile and other countries that are beginning to develop an environmental-policy framework and the institutional structure to implement it have enormous challenges ahead. But they also have important advantages over countries that began the effort some time ago. Chile can learn from the experiences of other countries and avoid making the same mistakes. Also, Chile has none of the inertia that has impeded change and improvement in those other countries. So it is important to introduce, from the beginning, the right concepts to avoid confronting strong inertial forces from bureaucracy, the industry, and the public.

CHAPTER 3[1]

ENVIRONMENTAL POLICIES AND PRACTICES IN CHILEAN MINING

Gustavo Lagos and Patricio Velasco

Mining, especially copper mining, has been important to Chile's economic development since the arrival of the Spanish conquerors in the 16th century. Between the 1840s and the 1880s, Chile's share of the world's copper-mining production rose to about 50%. This share was eroded by the emerging flotation technology in North America and by had decreased to only 5%. However, this share has increased steadily since the late 1960s, to 10.8% in 1970, 13.8% in 1980, and 17.3% in 1990; according to projections, it will be between 32 and 34% by 2000. But it unlikely to be more than 35–40% by 2010 (Minería y Desarrollo 1993).

The contribution of mining to the country's development has been of great importance. During the last decade, this activity has accounted for about 50% of the country's exports and foreign investment, about 5–7% of the gross national product, but less than 2% of the labour force. Copper is the main export, followed by gold, silver, molybdenum, iron, nitrates, iodine, and lithium. Some of the gold and silver and all of the molybdenum are produced as by-products of copper mining.

Most mining is concentrated in the northern provinces of Chile, one of the driest places in the world. This is a region with little agriculture, no forestry, and few towns. If one could select a region in the world where the environmental impacts of mining must be at a minimum, northern Chile would be at the top of the list. Most of the northern economy depends on mining — this activity may not be important in terms of employment at a national level, but it is paramount to most northern cities and towns.

[1] This project was carried out with the support of a grant from the John D. and Catherine T. MacArthur Foundation of the United States, coordinated by Alyson Warhurst.

The copper-mining boom of the 1990s is associated mainly with large, new, foreign-owned mining operations using the latest exploitation and processing technologies and the most modern environmental-management methods. By 2000, close to half of the Chilean copper production will come from these new mining projects, and the remainder will come from mines that have been active for several decades.

The most important old copper mines are Los Bronces and El Soldado, owned by Exxon Minerals Co. through Compañía Minera Disputada de Las Condes; the Mantos Blancos mine, owned mainly by Anglo American Corp.; and the Chuquicamata, El Salvador, Andina, and El Teniente mines, owned by the state company, Corporación National del Cobre (CODELCO, national copper corporation). It should be added that CODELCO's mines were nationalized in 1971, after belonging to companies from the United States. CODELCO's production constituted 85% of Chilean copper production in 1980; 75%, in 1990. No foreign companies remained in Chile's mining after the industry's nationalization process. However, in 1974, General Pinochet's military government issued foreign-investment decree 600, which gave rights to foreign companies to make certain uses of their investments and revenues from these investments. In 1983, Chile issued a new mining law. This law, which had constitutional status, established methods for calculating compensation for mining properties expropriated by the state.

The mining boom of the 1990s has been due mainly to foreign investment. Chile's attractiveness to foreign investors lies not only in the exceptional quality of the country's ore deposits and its very favourable geography but also on the laws decreed in the previous decade, Chile's economic and political stability, and the quality of its labour force.

As will be demonstrated in this paper, two main trends can be identified in corporate environmental policies and practices. These policies and practices originated with state-owned companies and with foreign companies that entered Chile's mining industry after the nationalization process. In addition to CODELCO, a second state-owned mining company is La Empresa Nacional de Minería (ENAMI, national mining company). ENAMI's role is to buy and process copper and gold minerals, concentrates, and precipitates from small- and medium-scale mining companies. ENAMI owns and operates five mineral-processing plants and two smelters.

The environmental impacts of mining in Chile can be traced back to colonial times but became more apparent during the late 1970s and especially the 1980s. This was due to the growth of mining production, impoverishment of ore grades, and increasing metallic impurities in tailings from processing plants and smelter-feed material. The public's awareness of and international concern about environmental impacts also grew during those decades.

Companies made no assessments of environmental degradation before the late 1980s. A study published during this period (Lagos 1989) pointed out that the main environmental impact of mining in the 1980s was air contamination produced by sulfur and particle emissions from the six copper smelters in Chile at that time. This was possibly followed by river and sea contamination from tailings dams, leaching plants, smelters, and even natural waters that passed through open-pit or closed-pit mines, picking up metal ions and, in some cases, suffering changes in pH (Lagos 1990).

These hypotheses were in fact validated by President Aylwin's government, which came to power in 1990. This government made its greatest efforts in the mining sector in 1990 and 1991, dictating decree 185 for the regulation and control of emissions of SO_2 and particles from fixed sources (mainly from the seven copper smelters that have operated in the 1990s). Decree 185 establishes the same air-quality standards for northern Chile as the US *Clean Air Act* establishes for the United States. For central and southern Chile, where there is more rain, decree 185 establishes more stringent standards because of acid-rain effects.

Despite its economic importance, mining is not the main contributor to environmental degradation in Chile. More important factors are desertification, erosion, urban growth, and possibly industrial activity and forestry exploitation.

The object of this paper is to establish the main environmental impacts of Chilean mining since the 1980s, its relative importance at local and national levels, and the trends in company environmental policies and practices. It analyzes the strengths and weaknesses of those policies in the context of the existing legislative and institutional framework. For this purpose, we chose four mining companies as being representative of corporate environmental policies and practices:

- Two state-run operations — CODELCO's El Teniente mine and ENAMI's Ventanas smelter;

- Compañía Minera Disputada de Las Condes, which was bought by Exxon Minerals in 1978; and

- Compañía Minera El Indio gold–copper mine, which began its operations in 1980 and was owned then by the St Joe Mineral Company, from the United States.

In many respects, these mining operations are quite diverse: the El Teniente and Disputada mines are comprehensive copper operations — from mine to smelter; the Ventanas works is a smelter plus electrorefinery and noble-metals plant; and the El Indio works is a gold mine, whose products are doré metal and copper concentrates. These differences do not detract from the value of the study because the idea is not to make a quantitative comparison of the processes involved but to examine the evolution of corporate environmental policies and practices. Thus, the results are qualitative and may help to enlighten future issues in the mining-and-environment scenario in Chile.

Environmental impacts of mining in Chile, 1980–94

Environmental impacts of copper smelters

Seven copper smelters operate in Chile; their environmental impacts can be summarized as follows (Solari and Lagos 1991; Schwarze and Muñoz 1991).

The Chuquicamata smelter, owned by CODELCO, produced 420 000 t of blister copper in 1993 and is the largest copper smelter in the country. It is located in a desert. Its emissions have diminished considerably since the mid-1980s, when four new acid plants were constructed and equipment to fix particles was introduced. When the wind blows in a westerly direction, the smelter impacts mainly on the mining camp of Chuquicamata, where workers' families lived until recently. The town has been almost completely evacuated because of this pollution.

Contamination from the smelter has no direct affect on Calama, a town of several tens of thousands of people about 16 km southwest of the smelter. The emissions include SO_2 and particles containing arsenic, copper, and other elements. In 1993, about 75% of the arsenic and 50% of the sulfur were being fixed. Plans were to reduce the sulfur and arsenic emissions to comply with decree 185 by the end of the decade. Chuquicamata has the natural advantage of having the wind blow generally in an easterly direction, over a desert region.

The Potrerillos smelter (also owned by CODELCO) has the same type of advantage. It produced 135 000 t of blister copper in 1993. The Potrerillos smelter impacts on the mining town of the same name, where workers' families live. However, it is understood that reduction of the smelter's fugitive gas emissions will substantially alleviate, at very low cost, the impact on Potrerillos town, although the total gas emissions through the stack will remain constant. To comply with decree 185, however, this smelter should install two acid plants.

A third CODELCO smelter is at Caletones, located in a high mountain area 113 km south of Santiago. This has at present only one small acid plant that only fixes about 2.5% of the sulfur in the smelted ore. In 1993, the smelter produced

364 000 t of blister copper. Its impact occurs mainly over the Caletones area, where workers operate the smelter, and also to the west over Coya, a now semi-abandoned valley mining town with agricultural land. This smelter has a lower impact on Rancagua and other towns in the central valley of Chile, including Santiago, where the stringent standards set by decree 185 apply. The arsenic effects may be of more relevance, but in 1991, the smelter reduced its use of feed concentrates containing arsenic. Caletones was to get a new acid plant in 1997 that will fix about 40% of the sulfur that goes into the smelter.

ENAMI, the other state-owned mining company, owns and operates two smelters, Hernán Videla Lira (also known as Paipote) and Ventanas. Paipote is near the city of Copiapó, in northern Chile. its emissions impact mainly on the nearby agricultural valley. In 1993, this smelter fixed between 25 and 30% of the sulfur, but this value will improve as more of the capacity of the acid plant begins to be used. Indeed, ENAMI plans to install a second acid plant at Paipote to comply with decree 185 by the end of the decade.

The Ventanas smelter is located on the coast, in the central region of Chile, and its emissions impact on the adjacent agricultural valley of Puchuncaví. Its acid plant, which began operating in 1991, can fix about 50% of its sulfur. ENAMI has considered a number of options to bring this smelter into compliance with decree 185. One of the most favoured options, considered in 1994, is to reduce the smelting capacity, which could also be a profitable move. In 1993, Paipote and Ventanas produced 63 000 and 140 000 t of blister copper, respectively.

All of these acid-plant projects have a negative profitability for the companies involved. The dilemma for the government is to find the resources to keep these plants operating when there is a need for social projects at the national level. The cases of CODELCO and ENAMI are different because ENAMI is a service company, without which small- and medium-scale mining companies, which make an important contribution to employment in northern Chile, would be economically viable. ENAMI's effective profit is very small because a significant portion is used to subsidize small mining operations, and the state is reluctant to spend its limited investment funds on projects that may not even pay for themselves.

A sixth copper smelter is Chagres, which belongs to Disputada and produced 75 000 t of blister copper in 1993. This was the only smelter in Chile that complied with environmental regulations in the 1980s and early 1990s, despite having possibly the "worst" location from an environmental point of view, being surrounded by fertile agricultural land and near a town. It should be added that in 1993 this smelter stopped operating for 6 weeks to comply with decree 185. It is being expanded to produce 155 000 t of blister copper per year and will change its present technology, adding a flash smelter. The emissions will be held constant to comply with decree 185.

A seventh smelter, belonging to the Chilean company Fundición Refimet S.A., started operating in August 1993 with a nominal capacity of 95 000 t of blister copper per year. Despite its use of reverberatory-furnace–converter technology, it was designed to comply with decree 185. The smelter was scheduled to expand its capacity to 125 000 t of blister copper per year in 1996 and then to 190 000 t in 1998. It plans to comply with decree 185 at all stages of its expansion.

Chile has improved its sulfur emissions considerably since 1989, when its copper smelters emitted 922 000 t of sulfur into the atmosphere and caused Chile to be ranked just behind the United States, China, and the former Soviet Union in sulfur emissions from fixed sources. Chile also ranked at the same level as Italy and Germany. In 1991, Chilean copper smelters emitted 15% less sulfur than in 1989. If Chile makes all of its planned investments in acid plants, it will reduce its sulfur emissions to 250 000 t by 2000, despite increasing its copper-smelting capacity by 27% over that of 1990. The investment ENAMI and CODELCO require to achieve this reduction is estimated to be close to 1 billion United States dollars (USD).

If the cost of compliance with the soon to be determined air-quality standards for arsenic is added into the equation, this cost could rise considerably (ACS 1991; Minería y Desarrollo 1992). This is especially true of Chuquicamata, whose ore has a high arsenic content. Chile is perhaps one of the countries most exposed to arsenic in the environment because of its very high levels of volcanic activity. Northern Chile has had a long history of dealing with arsenic: several rivers have base levels that greatly exceed the standard. Several plants extract arsenic from this water, but this has nothing to do with mining. The arsenic emitted into the atmosphere from some smelters and roasters in Chile only adds to this initially high base level in the water. In some cases, as at the town of Calama (located near the Chuquicamata smelter), the base level in the air is also high because the wind carries particles from the desert at a level that exceeds the air standard for arsenic in Europe and the United States. The high arsenic content of the desert seems to be surficial in some areas, which may be due to arsenic particles' being deposited on the surface for many years; it may also be due to the frequent and recent volcanic eruptions, such as the Láscar eruption in 1992.

The government has started a research project to determine the air-quality standard for arsenic. It is thought that the available world database relating cancer to total arsenic exposure could be improved by looking at Chile, as its environment has a natural abundance of arsenic. This research makes economic sense. For instance, if the standard for arsenic in the air is fixed at the levels proposed by the Ministry of Health, then the Chuquicamata smelter, and maybe also the mine, will have to close down. If, however, the standard is fixed at the levels proposed by international consultants, that is, at levels based on North American and European risk information, Chuquicamata will be able to continue operating. Therefore, the total acceptable exposure to arsenic is an important issue that has to be resolved in this decade.

Impacts on rivers

Studies by IRM (1988), Muñoz and Lagos (1990), Luna and Lagos (1990, 1992), and DGA (1991) of the environmental impact of mining activity on four important Chilean rivers — the Mapocho, the Elqui, the Aconcagua, and the Copiapó — showed that mining had had no proven impacts on agricultural or populated areas. However, the water-quality standards (stated in decree 1333 of 1978) were surpassed at certain points in these rivers during certain periods, as a result of mine effluent. These studies were based mostly on data collected in the mid-1970s by one of the state agencies, Direccion General de Aguas (DGA, general directorate of waters). These and other studies (Dames and Moore Consultants 1991; Lagos 1993) showed that the base levels in some parts of these rivers surpassed the water-quality standards of decree 1333.

The case of the Salado river was different. This river received tailings from the mineral-processing plant of the El Salvador mine, owned by CODELCO. The tailings completely embanked the bay of Chañaral and caused the death of sea species in the surrounding area. The El Salvador tailings dam had been filled since 1938, and the company, which then belonged to a US firm, allowed the tailings to spill into the Salado river. In 1975, the company, which was by that time owned by CODELCO, constructed a canal to divert the river water, including tailings, to another bay, Caleta Palitos, but this action produced the same embankment effect there. Overall, the company dumped 330×10^6 t of tailings into the river and the canal before the it was forced to change its tailings-management practices in 1989. In the late 1980s, an environmental group from Chañaral took CODELCO to court. The court in Copiapó ordered the company to construct a new tailings dam. In 1989, the Supreme Court ratified the decision, and CODELCO was forced to build a tailings dam, which has now entered into operation. This was an important precedent in Chilean law: despite the outdated environmental laws covering liquid effluent, companies, even state-owned companies, could be taken to court and forced to deal with environmental problems. This case dealt with the most serious environmental impact of mining on river or sea waters in Chilean history, and it illustrates clearly that environmental policy was not a priority for CODELCO until very recently.

Tailings dams

In 1993, Chilean mines had 717 tailings dams and 149 solid-waste deposits, according to studies by Servicio Nacional de Geología y Minería (SERNAGEOMIN, national geological and mining service) (1990) and Galaz (1993). Although these studies included only the northern region, the central region, and a small part of the southern region, most mining activity is concentrated in these regions. The conditions of 299 dams were found to be unsafe, meaning that either the walls were unstable or liquid was filtering out to surficial waterways or down to groundwater. Some of these dams were also located in dangerous areas where there was a risk of earthquakes or flooding. An undetermined number of the dams were abandoned.

Legislation on tailings dams dates back to 1970 but is insufficient for today's safety and environmental requirements. For example, this legislation gives no guidance on the construction of dams in waterways, which could be very dangerous in the event of earthquakes or flooding. Chile has had many episodes of earthquakes at tailings dams. The most important one was in 1965, when the El Cobre tailings dam, which now belongs to Disputada but was owned by ENAMI at the time, collapsed during an earthquake and killed more than 220 people in a nearby town. The collapse was attributed to the unsafe condition of the dam walls. Disputada was forced to introduce safety measures but was never compelled to pay any compensation to the families of the people who died. Chilean law has no precedent for a mining company's having to pay compensation to an individual or group of individuals for environmental damage. This will certainly change now that environmental-framework legislation (EFL) was approved by Congress in January 1994. The EFL defines the concept of responsibility for environmental damage, among other elements. In the past, the courts at most ordered companies to repair their facilities, install new measuring devices and treatment equipment, or temporarily close their plants.

The question of tailings dams located in waterways will be discussed in the subsection providing background for the case study "Compañía Minera Disputada de Las Condes."

The first mine to apply the new dam-design concepts in Chile was El Teniente (another of our case studies), with the construction of the Carén dam. The new concepts are based on several principles:

- The dam must be located where there is little likelihood of being affected by flooding;

- Any water eventually filtering from the dam should not pollute the groundwater;

- The dam should be located where there is little risk of earthquakes;

- Water from the dam should be channeled to agricultural or forest soils or to waterways able to withstand effluent without the quality of their water being affected; and

- The company should have plans to reclaim the land after the mine is abandoned.

These design concepts have now been extended to several other dams in Chile and are accepted as standard. Many companies in Chile are now reclaiming the land. In some cases, these companies have improved the conditions of the region, such as by growing forests where there was arid land before.

SERNAGEOMIN has been working to elaborate new legislation on tailings dams that incorporates these concepts and introduces several clauses on obtaining information about the movement of solid materials during mining. Leaching installations are included in this legislation, as well as the disposal of wastes removed from a mine when the mineral is extracted. This legislation will be an important improvement, but many aspects will still be insufficiently unregulated. One of these aspects, the reclamation of land at abandoned mines, is usually associated with abandonment of tailings dams, leaching dumps, and other installations.

Other aspects inadequately addressed by Chilean legislation, even after the EFL, are who pays for the reparation of environmental damage or for the reclamation of land abandoned by a mining company and what safety and environmental conditions are unacceptable. These problems will be discussed, again in "Compañía Minera Disputada de Las Condes," under the heading "The case of the Perez Caldera tailings dam."

Chilean legislation also fails to address acidification of water caused by mine operations. This problem is related to the operation and abandonment of mines, tailings dams, and leaching operations and is affected by natural water behaviour, by precipitation, and by atmospheric pollution from smelters and other fixed and mobile sources of sulfur. This is yet another subject that needs to be investigated soon before future legislation is written.

The soil

Not very much is known about the degradation of soil resulting from metal ions or other waste products from mines and mining plants; in fact, Chile has no soil standards. Many mines and mining plants in Chile are located in high mountain areas or in deserts, so if contaminants have accumulated in the soil, the impacts have gone unnoticed. Only a few cases of contamination of soil by river water polluted by mines have been reported in the last decade. One of the reasons for this may be that agricultural soil in Chile is usually neutral or basic, which can buffer any excess acid caused by a high metallic content in water; the metal precipitates, and no harm occurs.

In Puchuncaví valley, which is near the Ventanas smelter, researchers carried out studies to establish the effects of the emissions from the smelter and from the Chilgener thermoelectric plant on the agricultural land close by (Chiang et al. 1985; Gonzalez and Bergqvist 1985; Gonzalez 1992). However, these studies were unable to show a conclusive relationship, for example, between pollution and reduced soil productivity (Lagos and Ibañez 1993). Similar results were found in other agricultural regions near smelters. Thus, no farmers' lawsuits against copper smelters have been successful. There may have been some out-of-court agreements, but these have not been reported publicly.

The use of water in northern Chile

The lack of water in the north may be a matter of future importance, especially given the rapid growth of mining in the 1980s and 1990s. The effects of water use during mining operations in deserts are not very well understood. How does this affect the provision of water to the Altiplano, the very fragile and unique high-altitude plateau ecosystem of many regions of the Andes mountains?

It seems that many mining operations could use water more efficiently than they do now. DGA, an agency of the Ministry of Public Construction, has monitored and created sophisticated models of the supply of water to some of the main Chilean rivers. A new mining operation would not be granted water rights if a model indicates, for instance, that a particular ecosystem would not be able to withstand that use. One of the problems is that DGA still has no models for many ecosystems in the north, especially models of the behaviour of groundwater. Thus, a company that finds groundwater, usually after having investing a considerable sum of money in exploration, can be granted permits to use this water without anyone knowing in advance what the impact of such use will be on the ecosystem. The way the system usually operates is that the company must monitor water levels and other indicators of life in the ecosystem and provide these data to the authorities so that a model can be elaborated in the future.

Lately, Congress has been developing legislation to regulate the use of water. However, some of these clauses may conflict with decree 1333, which sets water-quality standards. The proposed legislation states that in northern Chile, all industrial and mining companies may dispose of used water in surficial waters or groundwaters provided this does not alter the base quality of these waters.

The proposed legislation considers another aspect, the duration of water rights. The proposal is to limit the duration of a new water-use permit granted to a company, even if that company did the exploration to find that water source. When the permit expires, DGA would acquire the rights to use that particular water source as it sees fit. An alternative proposal is that these new water permits be permanent, like the old ones, provided that a fee for the water rights is paid by the company. At present, water rights are permanent and require no payment. This means that a few large companies that have operated for many years have concentrated most of the water rights in Chile and can use this water under authorization of the DGA.

Congress will have to approve coherent legislation on these issues within the next few years, but the potential impacts of water use in the arid regions of northern Chile will remain unknown until much more is known about the behaviour of the affected ecosystems.

Public policy, regulations, and institutional issues

The Comision Nacional del Medio Ambiente (CONAMA, national commission of the environment) has identified about 2 200 laws and presidential decrees dealing with the environment, and many of these might be applicable to the mining sector (Lagos et al. 1991; CONAMA 1992; Bórquez 1993). However, some date back to the beginning of this century and are no longer useful. Furthermore, many of these regulations have never been applied. The Aylwin government began to use only a fraction of them, which shows that the application of environmental legislation was, and still is, discretionary, depending more on political will and state officials' readiness to act than on scientific data.

The main institutions and agencies responsible for the environment belong to different ministries, and they elaborated most of the current environmental legislation without the benefit of overall coordination. Thus, this legislation reflects to a large extent the specific interests of each ministry, rather than an overall concept of the environment. As a result, the rights and responsibilities of the various agencies often conflict. For instance, seven different institutions control the quality of water resources, with intersecting responsibilities and rights. This has created conflicts on many occasions, which not only reduces the possibility of protecting the environment but also delays the processing of the permits needed by the mining companies. In addition, a lack of funds has impaired the procedures and methods of the agencies responsible for the environment; staffing has been insufficient; personnel are often improperly trained; centralization has been excessive; the delimitation of functions is not clear; and methods for reaching decisions have not been specified.

This situation is a product of many years of disregard for the environment. The military government (1973–90) had no environmental policy other than ignoring the environment as much as possible. It is not by chance that in the mid-1980s, an influential economic adviser to the government, who later become the minister of finance, told a meeting of North American business people that they would find investing in Chile to be advantageous because they would not have to comply with any environmental law.

The public companies were among the worst offenders against the environment in the 1980s, and the institutions responsible for the environment made no attempt to change corporate behaviour in this period. A national commission of ecology was created, but it had no budget or the power to do anything except keep a few good, willing citizens occupied.

Parallel to this, public awareness about the environment began to rise. Legal demands were being presented in various parts of the country. President Aylwin's government (1990–94) based its policy on a program elaborated in 1988. This program called for the harmonization of economic growth with environmental protection (CPD 1989).

Since 1990, many actions have given the environment a higher profile in Chile. Among these are the following:

- The creation of CONAMA and the commission for the decontamination of the metropolitan region of Santiago;

- The involvement of regional authorities in environmental decision-making;

- The dictation of decrees 4 and 185 (the former is equivalent to the latter but is for the Santiago region);

- The elaboration of the EFL (which was approved by Congress in January 1994); and

- The allocation of funds to solve some of the environmental problems of the state-owned companies.

The EFL is of special importance because it is the first attempt to provide a framework for environmental regulations in Chile and to refer these to a coherent environmental concept. It gives CONAMA a new role: coordinating and eventually approving all environmental-impact studies (EISs). All projects of a certain size undertaken by a company, institution, or individual must first be subjected to an EIS. When CONAMA approves an EIS, it issues an environmental-impact declaration (EID). The EID sets down the specific conditions that the project must meet. CONAMA acts in each region through a Comisión Regional del Medio Ambiente (COREMA, regional commission of the environment), which actually determines the specific terms of reference for each EIS and later analyzes its results and may issue an EID. Thus, the COREMAs for each region and CONAMA, at the national level, constitute a single-window facility to coordinate the views of all agencies and ministries involved in environmental matters.

CONAMA is institutionally located within the Ministry of the Presidency, which coordinates all matters of the Presidency with other ministries. Nothing of state importance should by-pass this ministry. CONAMA has a directing body of nine ministers, headed by the Minister of the Presidency.

Another concept introduced by the EFL is the definition of responsibility for environmental damage. Any citizen, institution, or company may be found responsible for environmental damage if a plaintiff can prove that such damage did occur. The EFL also establishes the concept of *who contaminates, must pay*, meaning that other decrees and laws can define market mechanisms for environmental control. These concepts are already included in decrees 4 and 184, which pertain to the atmospheric emissions of fixed sources.

Despite the will for environmental action shown by the Aylwin government, it can be observed that the advance has been very heterogeneous. The mining sector has come out the winner, mainly because it went ahead to find its own environmental solutions without waiting for the institutional and legal changes to be completed. Instead, it worked to promote the ad hoc commissions, established through the goodwill of the different ministries and other institutions involved in environmental protection and with the cooperation of private and public companies. Contrary to what some believe, companies do want legislation and clear rules. In 1990–93, when the EFL was yet unapproved, most of the procedures for new projects — for example, procedures pertaining to EISs, baseline studies, liquid and solid effluent, and even abandonment — were agreed to without reference to any specific legislation, as such legislation was nonexistent or was too incoherent to be applied.

Such ad hoc commissions could be very useful in establishing realistic legislation in the future, but unfortunately the system has left a wide gap for state discretion, which, as history shows, is not always for the best in Chile. Even after the EFL, the gap is very wide, as the terms of reference are still a matter of discretion in cases to which no existing legislation applies.

One of the key aspects of approving the EFL was the opposing positions held by different political sectors in the country. Sectors favouring economic development believed that economic development should be the ruling consideration in the country's development, but other sectors believed that environmental protection should be the ruling consideration. Thus, the environmental advocates always pushed for the creation of an environment ministry that would have the last word on environmental issues (overruling other ministries), a more centralist vision than the EFL's concept of having environmental issues decided, in coordination, by all the institutions concerned. Had the environmentalists prevailed, the state apparatus would have had to have been completely restructured. The agencies that have historically had environmental control, however dispersed and incoherent, would have had to render their responsibilities to this new ministry. In the end, the state would have been reorganized with an environmentalist slant. But now state reorganization, which will go ahead during President Frei's government (1994–2000), will be geared toward improving the competitiveness of the country as a whole. But it is often argued that with this increase in competitiveness, the environment will be better taken care of after all.

The strength of the EFL is that it will be applied rapidly and, it is hoped, realistically; and the country's economic development may continue at the same rate as in the past 10 years (more than 6% average annual growth). Furthermore, the environment seems to be at its best in a competitive economy.

The EFL's main weakness seems to be that its optimistic view of the relationship between competitiveness and the environment may increase the risk of overlooking or dismissing some environmental-control measures, and this may lead to serious environmental problems.

Mining companies' environmental policies and strategies for environmental management

The following subsections outline case studies (Velasco and Lagos 1991) of the environmental policies and strategies of four mining companies: Compañía Minera El Indio, Compañía Minera Disputada de Las Condes, the Ventanas smelter (ENAMI), and the El Teniente mine (CODELCO).

Compañía Minera El Indio

Background

The El Indio deposit was discovered in the 1960s. The deposit had been exploited on a limited scale for about 10 years before the St Joe Gold Corp., from the United States, became interested in acquiring the rights to the deposit in 1975. This was the first foreign investment made in Chile under decree 600 (also called the foreign-investment statute), promulgated in 1974. In 1977, St Joe signed a foreign-investment contract of 100 million USD with the State of Chile.

In the 1980s, El Indio's property changed hands twice. In 1981, the Fluor Company bought a 90% share of St Joe. In 1987, Fluor sold its share to the Australian-based Alan Bond Group through its holding company, Dallhold Investments Pty. Because of financial difficulties in 1989, the Alan Bond Group sold El Indio to Lac Minerals Ltd, of Canada, which is still the sole owner.

The El Indio mine is located in the Andes at an altitude of 4 000 m above sea level (asl). It is 200 km east of the city of La Serena and is close to the Argentina border. The mine is exploited both by underground and open-pit mining. The extracted mineral is transported to a processing plant. This plant was designed with an original capacity of 1 250 t/d of dry mineral ore, but this was expanded, first in 1985, to 1 850 t/d, and then in 1988, to 2 600 t/d.

After the material is crushed, it is fed into a dissolver-drum reactor, where it is washed with water to remove the soluble copper and the ferrous salts. These are neutralized and eliminated as waste, as their presence would affect the flotation and cyanidation processes. During grinding, lime is added to the mineral, raising its pH to around 9.0, to prepare it for the flotation process. The copper concentrate obtained after flotation contains about 20% Cu, 50 g Au/t, 300 g Ag/t, and 8% As. This concentrate is then treated in a 14-level hearth-roaster furnace to remove the arsenic. The gas generated in the roasting process passes through two cyclones, to recover residual dust, and is mixed with air for the total oxidation of arsenic and gaseous sulfur. Continuing in the particulate-cleaning process, the gas passes through an electrostatic precipitator. It is then mixed with cool air, and a precipitate of arsenic trioxide is produced by lowering the temperature.

The tailings from this first stage undergo a second stage of flotation, followed by a pressure-filtering process, both to prepare the material for the recovery of silver and gold through cyanidation and to recover and reuse the water contained in the slurry. After cyanide leaching, the gold and silver undergo absorption with activated carbon and then desorption at high temperature and pressure. The rich solution is electrolyzed and then smelted to obtain doré metal with a 10 : 1 gold–silver ratio, by weight. The tailings are carried by gravity through a pipeline to the tailings dam; from there, water is pumped back to a reservoir at an altitude of 3 800 m asl.

On a few occasions in the past, cyanide-contaminated water has flowed from the tailings dam to the Malo river, one of the tributaries of the Elqui. The health service of IV Region took note of these events, and in 1982, a year after the cyanidation plant started operation, health officials closed it for 3 months. The plant was reopened after a process modification that reduced the cyanide content of the tailings dam from 600 mg/L to 10 mg/L.

The health service of IV Region periodically analyzes the level of harmful substances at different points in the river between the mining plant and the water-collection plant that treats the water and supplies it to La Serena. Maximum levels for certain substances have been decided for different points along the river. In Huanta, one of the first localities downstream of the mine, the water from the Turbio river (another tributary of the Elqui) must not contain more than 0.2 ppm of cyanide concentrates. A few kilometres downstream from Huanta, near Chapilca, the maximum allowable concentration of cyanide is 0.05 ppm. The third sampling point is farther downstream, and the fourth and last is in Las Rojas, near the water-collection plant that supplies La Serena. The maximum level of cyanide tolerated at both these sampling stations is 0.02 ppm. There has been no known incidence of these levels having been exceeded.

It is evident from analysis of DGA data that opening El Indio resulted in an increase in the contents of copper, arsenic, and cyanide in the Malo river, which passes close to the mine, plant, and tailings dam (Muñoz and Lagos 1990). Although it should be added that the Malo river (*malo* means bad in Spanish) has a base level of arsenic that is well above the Chilean standard for water (both for agricultural use and for drinking), it is also clear that none of the environmental events detected have had an impact on agricultural or populated areas. This is not only because the contaminants are diluted downstream but also because the soil is basic and precipitates the metal ions in the water on contact.

When the company was installing a third roaster in 1991, the health service of IV Region demanded that the company comply with certain requirements to control the arsenic its roasters released to the atmosphere, These requirements were, first, that by 1993 the arsenic emissions of the three roasters not exceed those produced by the two older roasters at the end of 1990; and second, that the company install a monitoring network.

Environmental policy

The environmental policy of El Indio is made according to guidelines established by its owner, Lac Minerals, of which the governing principle is sustainable development. From this broad definition stems a policy framework with the following seven strategic concepts:

- The company's administrative and operational practices must be compatible with legislation regarding the protection of workers, the community, and the environment.

- The company must have an autoinspection program to ensure compliance with government and corporate policies.

- When the company has been informed of developments in technology or processes that can reduce adverse effects on the environment, it should introduce these changes, if it is economically feasible, even if they exceed legal obligations.

- If no legislation exists, the company should apply effective procedures to promote environmental protection and minimize environmental risks.

- Environmental awareness should be promoted in the sphere of governmental authorities, workers, and the community to ensure that laws are fair, realistic, and economically feasible.

- The company should encourage and participate in research to find effective solutions to environmental problems, in harmony with cost-reduction strategies.

- The board of directors of the company, government authorities, and the community should be informed about compliance with environmental programs and responsibilities.

Environmental activities

The environmental activities of El Indio are separated into different areas, of which the most important are environmental management, surveillance, and control. Other areas include training, information, and coordination.

ENVIRONMENTAL MANAGEMENT AT EL INDIO — The environmental-management system is assessed by an environmental committee consisting of the general manager and the legal manager, the latter being most directly responsible for environmental and health management. Despite the existence of a central committee, in practice each individual operation area has to comply with specific environmental goals derived from company policy. The company, El Indio, as well as the parent company, Lac Minerals, periodically carries out environmental audits to determine whether environmental goals have been met and if necessary to introduce modifications.

Normative activities involve making the company's environmental policy public, with a view to demonstrating to government authorities and to the community the company's performance in the environmental-care sphere.

El Indio developed an environmental plan that takes all aspects and areas of the productive process into account. It determines critical points and priorities and compiles information on processes, installations, climate, topography, use of resources, and other issues. Using this information, the company developed an observance program that assigns responsibilities and establishes a periodic-assessment system and the budget needed to meet the goals. Environmental-protection recommendations have been developed for every component of the environment.

El Indio has also designed an environmental-emergency management plan to ensure coherent action to reduce adverse effects in a critical scenario.

A works-abandonment plan has been devised to cover the safe management of installations, deposits, and dumps after operations have ceased.

Finally, El Indio has compiled materials-security booklets for diverse environmental risks, and these booklets are distributed throughout the company.

INVESTMENTS AND ACTIONS OF EL INDIO — In environmental-surveillance and control activities, El Indio, in conjunction with regional entities (particularly La Serena university), has done a wildlife count in the zone to establish a baseline for local ecosystems. An EIS examining the interaction between mining and the environment complements this information.

A permanent monitoring system has been set up to control and survey the physical and chemical characteristics of the effluent in the operations area of the Elqui river basin, particularly the cyanide content. This surveillance allows the company to detect any change in water quality and make adjustments to the production process. The monitoring system's particular function is to feed a database with data about pollution in the Elqui river basin. It is worth noting that the industrial water used in the process is completely recovered to avoid polluting the watercourses. The Malo river was channeled around the operations area, thus avoiding contact with the tailings dam nearby. In addition, the dam has an infiltration pumping system that prevents underground contamination of any watercourse.

Because of the high arsenic content of the El Indio concentrates, the company has equipped the roasting furnaces with gas sets, which, through cyclones, cooling chambers, electrostatic precipitators, and bag filters, recover a considerable amount of the arsenic dust generated in roasting. According to company plans, dust recovery was to be 98% by the end of 1993.

Lastly, forestation activities have been carried out in areas where mining activities have altered the physical characteristics, and the company has plans to plant trees on dry tailings deposits to improve their stability.

Activities related to environmental protection, as well as to industrial hygiene, have resulted in operational investments of 64.6 million USD, about 20% of El Indio's total investments in all mining activities in the 1981–91 period. In 1992, El Indio assigned a total of 679 000 USD to environmental activities, which represents 7.4% of its annual budget; 54% of this sum was spent on improving the tailings dam.

Compañía Minera Disputada de Las Condes

At present, Disputada has two mines, Los Bronces and El Soldado; two processing plants, Las Tórtolas and El Cobre; four tailings dams, Perez Caldera 1 and 2, Las Tórtolas, and El Cobre; and one smelter, Chagres. The San Francisco processing plant was dismantled after the Las Tórtolas plant started in 1993.

Background

The Compañía Minera Disputada de Las Condes dates back to 1916, when the Disputada and San Francisco mines were first worked. In 1936, the Los Bronces mine, located in the Andes, 62 km east of Santiago and at an altitude of 3 400 m asl, was acquired by the Disputada company. During the early years, the extremely high copper grades (10–25%) produced from these mines allowed the company to use rudimentary technology, with low productivity. However, the high-grade veins started to become exhausted, and after the grade fell to 4–5%, the company had to introduce both mechanical methods and a concentration plant. Between 1920 and 1925, the company established the Pommerantz level, which was used as a transport level; the Perez Caldera benefitiation plant; and two cable lifts, one of 7 km, to transport the ore from the Los Bronces mine to the concentration plant, and one of 21 km, to move the concentrate from the plant to Las Condes county in Santiago.

In 1952, the company was sold to a French firm, La Société minière et métallurgique de Peñaroya, which acquired 87% of the shares; the other 13% remained in the hands of minor shareholders. In 1958, when the 4–5% grade veins were depleted, new installations were constructed; the most important of these was an underground concentration plant with a capacity of 346 t/d. Because of a further fall in the ore grade, Disputada at Los Bronces was forced to completely renew its installations to raise production to profitable levels. From 700 t/d in 1959, processing increased to 3 000 t/d in 1962, 4 800 t/d in 1978, 8 400 t/d in 1981, 12 000 t/d in 1988, and 37 000 t/d in 1991. Several facilities for 55 000 t/d (nearly 200 000 t Cu/year) were designed in the most recent expansion. Favourable market conditions in the second half of the 1990s may lead to a further expansion of all facilities to this capacity.

ENAMI acquired Disputada in 1972, within the framework of nationalization, for 5 million USD. The company was sold for 114 million USD to Exxon Minerals Chile, an Exxon Ltd subsidiary, in February 1978.

The Los Bronces mine treats its minerals at the nearby San Francisco mineral-processing plant. The tailings are sent through a tunnel to the new Las Tórtolas tailings dam located 37 km away in the valley north of Santiago. Two of the four tailings dams, the Perez Caldera 1 and 2, are located near the San Francisco plant, in the San Francisco river basin. After 1993, all the tailings contained in these dams were to be sent through a new tunnel to the Las Tórtolas dam.

In addition to the Los Bronces mine and its related installations, Disputada also acquired the El Soldado mine, located in Nogales commune, in V Region, 135 north of Santiago, as well as the El Cobre concentration plant. The first mining-rights concession of this deposit dates back to 1842. This was transferred to the Sociedad de Minas de Catemu in 1899 and later to the Compañía Minera Du M'Zaita, in 1919. In 1975, it was acquired by Disputada and later, in 1978, these rights were transferred to Exxon. The Chagres smelter was also acquired by Disputada.

The El Soldado mine treats its minerals at the nearby mineral-processing plant of El Cobre, and the tailings are sent to El Cobre tailings dam 4. This dam was to be replaced by a new one, El Torito, closer to the plant.

The Chagres smelter, in the Catemu area of San Felipe province, north of Santiago, is in the middle of an agricultural valley. It is the only smelter in Chile that complies with environmental regulations. It was the first plant to install a monitoring network, during the 1980s, and the first to be regulated by a special decree (now replaced by decree 185). Some of the concentrates from El Soldado and Los Bronces are smelted in Chagres, to fill its capacity. Until 1992, the capacity of this smelter was about 50 000 t of copper. However, an expansion planned for 1995 would increase this to about 155 000 t of copper; the sulfur emissions were to be maintained at a constant level of 7 000 t/year.

Environmental policy

When Exxon's Disputada started its operations in Chile, it introduced an environmental-policy framework, in accordance with the policies of its mother company, Exxon Minerals:

- To comply with environmental regulations or, if such regulations do not exist, to apply responsible standards;

- To prevent incidents and to design, run, and maintain installations with this purpose;

- To react quickly and effectively to incidents resulting from mining operations;

- To promote the development of appropriate environmental laws and regulations;

- To carry out and promote research on the impacts of operations on the environment, to improve environmental-protection methods, and to increase the capacity to make operations and products compatible with the environment; and

- To audit operations to ensure they comply with this environmental policy.

Disputada's environmental policy addresses three main areas of concern: compliance with the legal norms in the environmental field; prevention of incidents in works and reduction of impacts; and compatibility of the production process with environmental care.

When Exxon acquired Disputada in 1978, it perceived that Chile's environmental regulations lacked a coherent structure for considering environmental aspects globally. As a consequence of this initial diagnosis, Disputada's environmental policy aimed at complying with foreign standards in cases where Chilean legislation did not cover certain matters. This explains the emphasis the company placed on promoting the development of environmental laws and regulations. The lack of comprehensive legislation on some environmental aspects (such as a framework law) generates uncertainty, as current practices may change.

Guided by its policy, the company was able to react quickly to a report (known to many people in Santiago) that the Perez Caldera 2 tailings dam might collapse in an earthquake. Disputada looked at ways to reduce the risk, realizing that if the dam collapsed it would have serious implications, such as material losses or loss of lives. In addition, such an event would damage the company's public image.

To put its policy guidelines into practice, the company created an environmental-management body. Its job was to develop plans and models aimed at tackling environmental issues and to propose action in the three main areas of concern. The organizations that operate at the different facilities implement corporate environmental plans. The environmental-management body gives guidance on environmental matters and assesses the management of operations. Similarly, it channels information to and from operational organizations, keeping them up to date on local, national, and international environmental affairs that might affect the company's activities in any way.

The environmental-management body has two other spheres of activity: first, it concerns itself with the internal structure of the company; and second, it acts as a counterpart to government departments. In the domestic sphere, it has to inform management of risk situations, as well as establishing beforehand possible sources of conflict, with a view to defining strategies for action. In addition, it has to present an environmental-performance evaluation to the company.

In the external sphere, the environmental-management body has to interact with state institutions linked to environmental administration, to obtain the authorizations needed to run the works. This activity is particularly important when introducing modifications to the process, especially expansions of the production capacity or the construction of related installations.

Environmental activities

Environmental-management activities occur at two levels: domestic and external. At the domestic level, the most important activities are the following:

- Personnel training — It is possible to develop awareness of the importance of environmental care. In practice, this allows the environmental variable to be introduced into project design.

- Environmental-impact analysis — By permanently monitoring the works and making base studies of water, air, and land quality, the company can determine the impacts of the production process on the environment. This activity also gives the company the information it needs to make decisions about introducing technology (emission-reduction and residual-treatment equipment) or modifying its operations to reduce contamination.

- Environmental-risk analysis — By identifying aspects that pose the greatest danger, the company is able to face any risk.

- Recovery of affected areas.

At the external level, Disputada's most important environmental-management activities are the following:

- Promoting adequate environmental practices, from a technical and economic viewpoint;

- Educating the national community about the environmental issues, directing public concern into more productive channels, and contributing to and benefiting from local scientific and technological development); and

- Publicizing the company's environmental experience to show that environmental protection is not only compatible with productive activity but also accompanied by clear economic benefits.

THE CASE OF THE PEREZ CALDERA TAILINGS DAM — Construction of the Perez Caldera 1 tailings dam was begun in 1936, at an altitude of 2 800 m asl, along the San Francisco river, which flows down to El Arrayán (on the outskirts of Santiago) and later joins the Mapocho river in Santiago. In 1978, when the dam had reached its full capacity, the company received approval to construct a new dam, the Perez Caldera 2, adjoining the first one but farther downstream on the San Francisco river. The Disputada company constructed a second San Francisco tunnel (the first one was in Perez Caldera 1) to cause the river water to flow parallel to the dam. In 1987, the company began construction of a third tunnel, the Ortiga, to provide an emergency outlet in case the San Francisco tunnel became blocked. Even before the completion of the Ortiga tunnel, however, the excessive spring thawing blocked the San Francisco tunnel, and the dam collected more than 300 000 m^3 of water in 5 d. The water level rose to a mere 45 cm from the top edge of the dam, threatening Santiago. Had there been an accident, the water would have rushed down the San Francisco river basin, from 2 800 m asl, to flood El Arrayán, at 1 000 m asl.

Whether or not, despite the precautions, the risks of such an event happening were real, the public began to think this tailings dam should not be there, as at any moment an earthquake or some other natural phenomenon might have catastrophic consequences for Santiago. However, Chile had authorized the construction of both dams and had no legislation to determine who should pay to repair this situation. Two options were open: evacuate the contents of the tailings dam or close the Los Bronces mine. By chance or by design, the company approved the Los Bronces expansion plan at this time, which meant the construction of a new tailings dam in Las Tórtolas, at the same altitude as Santiago. This opened the possibility of constructing a tunnel to evacuate the Perez Caldera tailings, although this was to be at a significant cost to the company.

The company rapidly came to an agreement with the community of El Arrayán, which was suing the company in the local court. The agreement stated that the Perez Caldera tailings would be evacuated within an agreed period, following the expansion of Los Bronces. However, the company never committed itself in that document to continuing with the expansion. The question of who would pay for the evacuation of the dam if the proposed expansion did not take place became all important. No precedent existed in Chilean law for this situation. Should the company pay the costs, regardless of its plans? What if this became economically so unfeasible that the company decided to close the mine? These questions have still not been resolved in Chilean law, even after the EFL. Who will pay for the repairs or rehabilitation of the many tailings dams that have been abandoned and no companies own them any more?

Investments and actions

Disputada has made environmental investments in its various operative units. In Los Bronces, the principal investments have focused on channeling and diverting the river water: constructing and repairing the San Francisco and Ortiga tunnels, for example, represented a total investment of 17 million USD. These works were followed by the conditioning of the tailings dam, at 13.8 million USD. Similarly, the company made investments in dam forestation at Los Bronces and in a water-treatment plant, among other things.

At the El Soldado deposit, the company has made its main environmental investments in the tailings dam, whereas at the Chagres smelter, it has invested in a sewage-water plant; an environmental-surveillance system (network for monitoring air quality); gas- and particulate-collection hoods in the furnace and converter and in the copper-matte and slag outlets; windbreaks at the concentrate-deposit site; a connection from the smokestack to the acid plant; an effluent-neutralization plant; a concentrate sweep and vacuum system; equipment for concentrate and rainwater collection and concentrate pelletization; a gas-breakdown plant (including electrostatic precipitators); and a smokestack extension (102 m). Finally, the hydrosowing-in-slag project and the gardens surrounding the works must also be counted as environmental investments.

The production-expansion project in Los Bronces also benefited from the construction of a new loading pier at the port of San Antonio, which cost the company 8.8 million USD. Disputada considers 6.2 million USD of that to be an environmental component because it includes sealed storage bins, a double-door system, and covered belt conveyers, which are also protected with curtains (to protect the fruit exports that are also shipped through this port). Disputada's largest investment in environmental protection, though, was the construction of the mining-ore pipeline to evacuate tailings from Los Bronces (in the high Andes) and deposit them in a valley Las Tórtolas, north of Santiago. This pipeline was designed after an EIS called for measures to minimize the effects of transportation on the environment. The Las Tórtolas tailings dam also includes the construction of walls, a water-collection and -recovery system, and the forestation and irrigation of the 700 ha property. The company's total investment in environmental protection in 1978–91 was 111 million USD.

The Ventanas smelter

Background

Planning for the Las Ventanas refinery and smelter project started in 1955, when the Empresa Nacional de Fundiciones (national smelting company) was created. It was decided that the smelter would be constructed at the beginning of 1956, and an electrolytic refinery was included to process all the copper coming from the small- and medium-scale mining operations. After assessing various alternatives, ENAMI decided to locate the new plant in an agricultural zone in the Ventanas area, near the port of Quintero, about 150 km from Santiago. Nonetheless, until 1960, the construction of this installation was limited to some preliminary work, such as earth movement, refilling, and road preparation.

In 1962, the smelter was inaugurated; and 2 years later, the electrorefinery plant was completed. The design capacity of the plant was 84 000 t of electrolytic copper. However, in 1967 an extension was made to increase the refining capacity to 112 000 t, and a modification in 1986 enabled the plant to produce 200 000 t annually, starting in 1987. In the same year, a study, originally initiated in 1963, was restarted; its objective was to look into the construction of a sulfuric acid plant. However, the company did not install such a plant until 1990. The smelter complex includes a mineral-storage facility, as well as the smelter itself, which has the sulfuric acid plant, a mercury- and arsenic-extraction plant, an effluent-treatment plant, an electrolytic refinery, and a noble-metals plant.

Environmental policy

Since its creation, ENAMI has had a double purpose: to obtain the highest profitability in mineral-processing, -smelting, and -refining processes and to provide social aid to the small- and medium-scale mining industry.

Until 1990, ENAMI did not have an environmental department, which explains the lack of an environmental policy or actions aimed at decreasing existing contamination levels at its plants. ENAMI's new administration created the Communications and Environmental Management Department to design and implement a global environmental policy, to help the company adapt to new legal requirements, and to respond to the growing public concern about environmental issues.

To convert its environmental policy into action, the administration created the positions of communications and environmental agents. These individuals are responsible for local environmental management at each of ENAMI's plants. Similarly, each production centre has an environmental committee, made up of interdisciplinary professional teams. These teams are responsible for detecting and analyzing critical environmental situations and proposing environmental-control measures.

One of the first actions of this managerial structure was to carry out a diagnosis to determine the environmental impacts at the company's production centres. One of the conclusions was that the company was slow at making decisions, which has meant the postponement, sometimes for years, of investments in environmental protection. Such was the case for the Ventanas sulfuric acid plant, which was initially conceived in the early 1960s but did not materialize until some 25 years later, in 1990.

ENAMI's environmental authorities brought up another relevant issue: poor maintenance of industrial equipment was having harmful effects on industrial hygiene.

The diagnosis also emphasized that even though there had not been a global environmental policy during the military regime, investments had been made to solve contamination problems, principally in the last years of the regime. However, companies had made those investments in response to public pressure, rather than on the basis of a clear environmental policy.

Investments and actions

Based on these observations, the Communications and Environmental Management Department began to design short-, medium-, and long-term environmental policies that took into account ENAMI's limited resources and harmonized the need for adequate production performance with the need to reduce contamination levels in a short time. The department came up with the Urgent Plan for Environmental Hygiene, which was intended to solve environmental problems in ENAMI's works relatively quickly and at a low cost period. The plan stated that the company's major concern was for the neighbouring communities, near the production centres.

ENAMI pointed out the importance of determining the real contamination levels attributable to the company on the basis of scientific observations before it implemented any initiative for the communities. Thus, it decided to carry out a study on the impacts of gases in the Ventanas zone during July 1990 and to compare its results with measurements taken at times when smelter production had been suspended. By taking into account other stationary sources that contribute to contamination, such as the Chilgener thermoelectric plant near the smelter, contamination base levels in the region could be established.

On the basis of this study, ENAMI established that the smelter emitted more SO_2 than the thermoelectric station, whereas the latter was responsible for more of the dust and particulate contamination.

A joint ENAMI–Chilgener – Ministry of Mines committee was created. Using the results of the study as a starting point, the committee was to propose a decontamination plan for Puchuncaví valley and a solution to the contamination problem in the zone. Within the framework of this plan, several environmental actions were carried out in 1991, at a total cost of 1.5 million USD. Most notably, the company

- Constructed a meteorological station to detect wind dynamics;

- Installed, in the Ventanas complex (and in ENAMI's other smelter), an air-quality-monitoring network consisting of four permanent measuring stations connected to a computer information-registration system (500 000 USD); and

- Hired professionals to conduct a feasibility study (500 000 USD) on modernizing the smelters, with a view to harmonizing smelting demands with environmental-control requirements.

These investments are included in the 57 million USD cost of the new sulfuric acid plant, inaugurated in 1990.

The decontamination plan for the Ventanas area aimed to have the smelter comply fully with decree 185 by 1999. ENAMI would invest 130 million USD to install a second sulfuric acid plant and to modernize some of the equipment. However, this plan changed several times in 1990–94, and the latest approach seems to be to minimize investment and maximize profits. This could be achieved by reducing production at the smelter, possibly by 1997. Also under consideration were the installation of electrofilters at the Ventanas smelter to collect arsenic dust and particles; and a forestation program for Puchuncaví, to be carried out jointly with the Corporación Nacional Forestal (national forest company).

El Teniente

Background

Large-scale production of the deposit known as El Teniente started at the beginning of the 20th century, when a North American, William Braden, bought the exploitation rights and set up the Rancagua Mine Company in 1904, also known as Braden Copper Co. Kennecott Copper Company became a partner in 1917 and later acquired the whole operation.

In 1968, with the so-called Chilienization process, the state bought 51% of the shares of the Great Copper Mining Industry. In 1971, a constitutional reform allowed the nationalization of the Great Copper Mining Industry. In 1976, CODELCO was created, uniting the four mines acquired through the nationalization process: Chuquicamata, El Salvador, Andina, and El Teniente.

The El Teniente mine is located in the Andes at an altitude of more than 2 000 m asl, 80 km southeast of Santiago. It is the largest underground mine in the world, with more than 2 000 km of tunnels and galleries. Close to 100 000 t of ore is extracted and treated daily. In 1991, El Teniente produced nearly 300 000 t of copper, mostly using the traditional process of flotation, smelting, and refining. Only 4 000 t was from electrowinning. El Teniente has a mineral-processing plant in Colón, a few kilometres from the mine; a second processing plant at Sewell, which is being phased out of operation; a smelter at Caletones, next to Sewell, a few kilometres south of Colón; and an electrowinning plant, also in Colón, to treat the mine water and the solution coming from an in-house leaching operation.

Environmental policy

CODELCO has had several projects with an environmental scope since 1976, but it wasn't until 1990 that the company began to design a corporate environmental policy and establish an environmental-management body. Its corporate policy is based on making environmental care compatible with competitiveness.

At El Teniente, the corporate policy was implemented through the creation of an environmental-control program, to be run by the manager of engineering. The first priority of this program was to establish a strategy for dealing with environmental problems at CODELCO's El Teniente division. To fulfil this goal, a diagnosis was undertaken to identify the division's main environmental problems.

The company envisages a research program to increase innovation in the production process and to apply the best environmental technology. This research program will promote the participation of all the executives, supervisors, and workers to preserve and improve the environmental resources affected by the division's operations. In addition, the program includes a diffusion policy to promote an accurate public perception of the realities of the environmental effects and to publicize the actions being undertaken to counter these effects.

Lastly, CODELCO plans to implement projects to improve environmental conditions. Tools include environmental audits, environmental-risk analyses, and emergency plans.

Investments and actions

The El Teniente division has already carried out or plans to carry out initiatives to reduce contamination levels at different stages of the production process.

The Carén reservoir, along with complementary installations, stands out as a major project. Established in 1986, it replaced the Barahona, Cauquenes, and Colihues tailing dams, which were already full. Located east of Rapel lake, 70 km from the Colón plant, the Carén dam was planned to have a 25-year life span, but this could be extended to 75 years. This tailings dam was the first in Chile to incorporate state-of-the-art environmental design concepts. The cost of this project was 189 million USD in 1986.

Clear water is evacuated from the dam into Carén and Alhue brooks, which flow into Rapel lake. This water can be used in agriculture, as has been demonstrated by experiments carried out by the El Teniente division at the Loncha experimental station, located next to the dam. Despite the high alkalinity of the water, it has been successfully used for horticulture and grain crops, as well as in forestry and aquaculture. The division also has an animal-breeding program, in which the animals are fed fodder and other products gown with this water at the Loncha station. According to the division, the experiments have shown that the water from the reservoir can be used as drinking water for animals and can increase the sustainability and efficiency of agriculture and forestry.

Other companies in Chile have emulated the construction of the Carén dam, together with its experimental station, in places where Chileans commonly use dam waters in agriculture, forestry, and even in cattle grazing.

It should be noted that the molybdenum and sulfate contents of the water from Carén dam are higher than the standards defined by decree 1333. At present, a special decree authorizes El Teniente to discharge this water. In the past 10 years, fish have died during two episodes at Scorpio bay in Rapel lake. The reports from the county authority indicated that these events were related to sodium hydroxide added to the sewage system to clear blocked pipes. Although other explanations were suggested at the time, these alternatives were far from being proven. In any case, the Carén dam will have to comply with the Chilean standards for sulfate content eventually. To achieve this, El Teniente has set up a research and development program with a view to constructing a plant to remove the sulfate and molybdenum from the water.

Another important environmental action the mine carried out in the last decade was the construction of a solvent-extraction and electrowinning plant in Colón. This plant treats the highly acidic waters that flow through the mine, which have a high metal content. Without this plant, untreated mine water would be discharged directly into the river. The investment in this plant was 25 million USD.

The third most expensive environmental project, at 19.5 million USD, was the handling system for smelter gases. This project was designed to protect workers from fugitive emissions in the smelter. Also included were the modernization of some available equipment and the installation of new equipment to reduce particle emissions. Despite these improvements, El Teniente's stack still emits more than 97% of the sulfur contained in the treated ore. It is therefore imperative that CODELCO construct acid plants at this mine soon to comply with regulations.

El Teniente has also been carrying out many less costly environmental projects related to different aspects of its very large operations. Most of these improvements do not have as great an impact as the three projects mentioned above, but the cost of carrying them out is less. One of these smaller projects is the forestation of the abandoned dams. The company has also identified more than 30 environmental projects that have yet to be done.

CODELCO's environmental investments in the El Teniente division between 1976 and 1990 were 240 million USD. The total investments in this division in the same period were about 1.157 billion USD, which implies that the environmental share exceeded 20%. However, CODELCO classified the total tailings-dam investment as environmental, whereas Disputada did not consider its investment in Las Tórtolas dam as environmental but as part of production expansion. In both cases, the dams were build because existing dams were filled to capacity. If one takes the total investment at El Teniente and subtracts all of the Carén investment, then the environmental share falls to about 5%. However, this figure ignores the fact that the Carén dam incorporates many state-of-the-art environmental design concepts.

Comparative analysis

The four mining companies have quite different histories. The processes employed are similar, except in the case of El Indio, which produces mainly gold, with copper as a by-product. Therefore, this analysis compares, not the environmental impacts of in each company, as these are very diverse, but the ways the companies face environmental problems, organize themselves to protect the environment, and interact with the emerging legislative process and changing institutions.

El Indio is the newest of the four companies and the only one that started up after environmental issues began to seem relevant. Thus, one can trace environmental policies at El Indio to its creation as a project. Ventanas dates back to the late 1960s; El Teniente and Disputada, to the beginning of the century, although one of Disputada's mines, El Soldado, dates back to the 1840s. Disputada's environmental history starts when Exxon took over from ENAMI at the end of the 1970s. Both El Indio and Disputada adopted the expertise of their US owners, as well as their code of ethics regarding the environment, and faced environmental legislation that was far from coherent. And both companies faced serious environmental problems.

When these two foreign companies entered the Chilean arena, they were confronted by a strong bias against foreign mining companies operating in the country. This bias was expressed mainly in some centre and leftist political sectors, including the trade unions. They argued that foreign mining companies failed to leave enough of their profits in Chile. This was the reason for nationalizing the copper mines 10 years before, and the view was still widely held in the 1980s. People openly disagreed with the new investment legislation dictated by the military government in the late 1970s and early 1980s. Despite the legislation, an anti-foreign-investment attitude also permeated the state apparatus. For instance, a special decree in 1985 compelled the Chagres smelter, owned by Disputada, to comply with air-quality regulations. Meanwhile, the five remaining copper smelters, belonging to the state-owned companies, CODELCO and ENAMI, carried on business as usual, without complying with any environmental regulation.

This bias still exists, but to a slightly lesser extent. Environmental organizations and the public have questioned most foreign-company EISs so far, whereas many EISs of the state-owned and Chilean companies have gone unnoticed. The foreign mining companies have reacted by adopting a policy of paying more attention to public concern about the environment. In addition, the foreign companies participate openly in the process of shaping future legislation. This is especially important because foreign mining companies still have to comply with more environmental requirements than state-owned companies do.

The analysis of the practices of Disputada and El Indio indicates that they took stricter environmental measures than required by Chilean legislation in the 1980s. The early compliance of Chagres is but one example. The expansion of the Los Bronces mine and the construction of the Las Tórtolas tailings dam, after the Perez Caldera tailings dam episode, were due partly to the need to evacuate the Perez Caldera tailings.

The advanced environmental-management practices of El Indio and Disputada were never adopted by CODELCO or ENAMI during the 1980s. Nevertheless, these practices served to show the Chilean companies how to approach environmental issues and helped the state apparatus shape future laws and regulations.

Disputada and El Indio openly discussed environmental issues with the public at a time when such practice was taboo at CODELCO and ENAMI. It was only after the Aylwin government came to power in 1990 that the state companies joined in the discussion of environmental legislation.

The two foreign companies probably had many motivations for adopting advanced environmental policies. First, their shareholders would not have let them get away with environmental degradation abroad when they were compelled to comply with strict regulations at home. Second, the managements of these companies knew that the global trend toward environmental regulation would sooner or later come to Chile and that it would pay off for them in the long run to take these advanced measures now. In fact, it had been demonstrated in developing countries that it is cheaper to install a clean process to start with than to pay for environmental repairs or a retrofit later on. Moreover, efficient technologies that allow lower production costs are almost always cleaner technologies too. (Because efficiency and cleanliness are integrated today, it is very difficult to calculate the environmental fraction of a new technology.)

The environmental policies of these two companies are much more elaborate in formulation and application than those of the two state-owned companies. This is not surprising, as CODELCO and ENAMI have had environmental policies only since 1990. Moreover, CODELCO's organization and priorities differ from one mine to another; El Teniente is one of the most advanced mines.

The strength of all four companies is that the environment is high on their lists of corporate concerns. Each company

- Examines new technology options;

- Trains personnel to handle environmental problems;

- Keeps its board informed;

- Prevents incidents and has emergency plans;

- Identifies priorities and advises and controls management;

- Undertakes environmental-risk analyses and does research on the environment; and

- Divulges the company policy.

Some of the strengths of El Indio and Disputada are the following:

- Their operations are currently compatible with environmental concerns;

- They promote environmental awareness to the government and to the public;

- They promote environmental laws;

- They have periodic audits, works-abandonment plans, and a decentralized environmental management; and

- They apply rules over and above Chile's environmental laws.

The state-owned Ventanas smelter and El Teniente mine are also involved in some of these areas, but they are not as yet official practice.

Disputada has displayed an ability to respond rapidly to events, as in the case of the Perez Caldera tailings dam. Also, by operating its Chagres smelter successfully according to legislation, Disputada has set the standard and method for other Chilean smelters. Regarding the Perez Caldera case, it should be remembered that construction began in the mid-1970s, when people had limited experience with the risks of building dams at high altitudes, either in Chile or in developed countries. Were one to apply the same standards today to all tailings dams in Chile, one would find many in the same situation.

Among the weaknesses of El Teniente and Ventanas are the following:

- They are unable to decide for themselves how much to invest on the environment because they have to compete for state funds that are in demand for much higher social priorities;

- The present management inherited environmental problems that will take many years to solve, even if the funds are made available; and

- The companies' structures and brief experience in environmental management will make it difficult to implement modern management practices.

The environmental-investment levels of the four companies were similar, at 18–20% of total investment. The problem is that comparing relative levels of investment does not necessarily show how willing the companies are to solve environmental problems but points to the diversity and complexity of the operations involved. More indicative than looking at past investment is to analyze present needs for environmental investment. Ventanas and El Teniente have still to invest very significantly in environmental projects, whereas Disputada and El Indio comply with regulations at present and are thus unlikely to have to make any new significant investments.

But where will the state companies get the funds they need to comply? The Ventanas smelter has found a way to comply with decree 185 that minimizes investment and maximizes profits, and at the time of writing it looked as if this smelter might be in compliance with decree 185 by 1997 or 1998 (1 or 2 years before the deadline), according to the decontamination plan. A first acid plant was projected to be in operation at the Caletones smelter by 1997–98. Whether this smelter will comply with decree 185 after this installation is still to be proven.

The state has a greater incentive today to authorize expenditures on new acid plants at Ventanas, Caletones, Paipote, and El Salvador, as Chilean public awareness is higher now and the free-trade treaty with the United States may be affected by such matters. If the state fails to authorize the funds required for these

installations, the only real option will be to sell the operations to private companies. However, the situation seems to be far from this as yet.

Conclusion

in the short term, the EFL is unlikely to fully resolve the inconsistencies of the
present legislation or to address the problems in Chile's environmental institutions.
The environmental policies of the various agencies and ministries still vary
greatly, and it will take many years to bring these into a coherent framework.
Nevertheless, the EFL represents a real advance in ordering the discussion of these
matters, in requiring EISs for all medium- and large-scale mining projects, and in
establishing the concept of responsibility for environmental damage, which did not
exist in Chilean law before. The EFL also means the effective decentralization of
environmental decisions, as the most important decisions will be shaped at the
regional level.

Legislation is still needed in this decade to deal with many environmental
aspects improperly treated or left unconsidered in the existing legislation. In the
meantime, the terms of reference of EISs and EIDs can be used to balance environmental, economic, social, and political interests. These decision-making tools
and their outcomes will be important to the design of future legislation.

It is evident that consensus agreements obtained in other spheres of Chile's
political, social, and economic development are not that common in terms of state
environmental policy. An important sector of the country believes that pressing the
country toward rapid economic development will have a considerable negative
impact on the Chilean environment. However, Chile is a democracy, and however
important this dissident sector may be, it will have to abide by the majority in
Congress, which so far has supported the EFL option.

One of the weaknesses of the EFL is that it fails to consider mechanisms
for community participation in environmental decisions. It should be remembered
that the community, acting through informal regional and local channels, has fundamentally influenced recent EISs. Excluding communities from participating in
decisions concerning EISs and other environmental issues may result in an even
larger dissident sector.

The policies and practices of mining companies in Chile seem to be way
ahead of the legislation and the institutional system. Even the difficulties of the
public companies lie not so much in their management's perceptions as in the
companies' histories and cultures, the volume of environmental problems they
face, and the lack of funds. Compliance with international environmental standards, especially those related to managerial practices, also depends on the transformation of the state-owned companies' vertical administrative structures into

lean, decentralized structures. The more experienced foreign mining companies may play a very important role in that transformation because they remain a show-piece of how environmental issues should be handled. The application of existing legislation and the design of future environmental legislation should also be influenced by the proactive practices of these foreign companies, especially given that the weight of the private mining sector is increasing faster than that of the state sector. It seems likely that compliance with managerial environmental stand-ards will be achieved before compliance with air-emission standards, as the former is not dependent on capital investment.

Despite the absence of legislation on many aspects of environmental management, the mining sector is putting its environmental policies into practice for all new mining projects. This has been possible because of the cooperation between government and industry. The examples are many, and one is led to conclude that new projects — like Candelaria, Quebrada Blanca, Cerro Colorado, Collahuasi, Zaldivar, Lince, Radomiro Tomic, El Abra, Manto Verde, and Refimet's smelter — have already met or will soon have to meet environmental requirements that are not too far behind those in North America, Europe, or Japan.

The participation of the mining industry was paramount to the elaboration of decree 185. The public companies need an up-to-date perception and knowledge of environmental problems not covered by the legislation. This would be the case for with solid- and liquid-effluent regulations, soil-quality standards, legislation on tailings dams, and abandonment procedures. Public companies must get involved in the study of these problems; otherwise, new legislation will not be firmly based on the possibility of these companies' achieving the new standards, as in the case of arsenic.

Public environmental policy was revolutionized in 1990. What occurred in CODELCO and ENAMI is an indication of this revolution. Other indications are the passage of decree 185 (regulating sulfur and particle emissions from fixed sources) and the discussion to set up the EFL. It is important to stress that pieces of legislation are being written, discussed, and approved with a sense of reality, for the aim is not to produce declarations of principle but to create effective legis-lation. In a country like Chile, the efficiency of such legislation is of paramount importance, as Chile cannot afford failed experiments — other sectors have such a great need for public investment.

CHAPTER 4[1]

ENVIRONMENTAL MANAGEMENT IN A HETEROGENEOUS MINING INDUSTRY: THE CASE OF PERU

Alfredo Núñez-Barriga, assisted by Isabel Castañeda-Hurtado

This research was conducted to examine the environmental problems of the mining industry in Peru, looking into its plausible explanatory factors. The research focuses on the development of legal and institutional framework, using an historical approach to place the sector in the context of the national economy, and the environmental behaviour of the mining firms, approached through a detailed analysis of case studies.

The analysis of the legal and institutional development framework aims at determining the extent to which its particular features have had a bearing on the environmental behaviour of the firms and have thereby limited the environmental impacts of the sector's activities. On the other hand, the analysis of case studies aims to work out the extent to which distinctive patterns of environmental behaviour may be ascribed to differentiated main mining groups.

This paper summarizes the main findings of the research.

The need to harmonize mining production and environmental control

This research aims to show, first, that the mining sector has traditionally played a key role in the Peruvian economy; and second, that it has also been a major contributor to environmental degradation in this century.

---

[1] The original study was developed within the framework of an international collaborative research project on environmental management in mining and mineral processing, centrally coordinated by A. Warhurst, Director, Mining and Environmental Research Network (see Warhurst 1991b). The specific terms of inquiry were those set forward for a Peruvian case study in Núñez (1991).

These hypotheses were kept in mind throughout the development of the research. In the end, we wanted to be able to find ways that improvements in the competitiveness of this sector — crucial to a developing-world economy such as Peru — could be harmonized with a regulatory system that could successfully stop the processes of environmental degradation.

Peru and the world mining industry

The importance of Peru in world mining goes back to the 16th century, when Spanish colonial rulers integrated Peru into the world economy. However, at the beginning of the 20th century, precious metals, fundamentally silver, gave way to the production of base metals: first came copper, then lead, and later zinc. Silver continued to be important but was produced mainly as a by-product of lead and zinc. Gold was important in the 1930–40s, declining afterwards and only returning to importance in the 1980s. Iron's importance as a main product was only acknowledged in the early 1950s. The major rise in gold production of the 1990s is being accompanied by the appearance of tin as the new important product of Peruvian mining.

Mining and the national economy

Mining has played a central role in the Peruvian economy as the main provider of foreign exchange. The median participation has been between 45 and 50% of total exports, and its share of gross domestic product has ranged between 9 and 10% in the past 2 decades.

However, mining is only a minor direct provider of employment; in 1989, it employed 1.3% of the working population. This figure may underestimate its actual capacity as an employment generator, because it fails to include the employment indirectly generated through mining's connections with other sectors of the economy. Nonetheless, even if this is included, it is unlikely that a qualitative change would be observed.

Mining and the environmental profile of the country

Documented information and interviews with officials and professionals working on environmental control in the country already show that the mining sector is greatly responsible for its past and current environmental degradation. Two of the more important institutions involved in this are the Oficina Nacional de Evaluación de Recursos Naturales (ONERN, national office for the evaluation of natural resources) and Direccion General de Salud Ambiental (DIGESA, general directorate for environmental health).

In 1986, ONERN produced, through international cooperation, the first attempt to diagnose the environmental situation in Peru (ONERN 1986). This work aimed to coherently integrate a large number of partial studies and information and to define a reference framework of priority areas and problems. Similarly, it also coordinated the production of an official national report (ONERN 1991) for the United Nations Conference on Environment and Development held in Rio de Janeiro in 1992. Both documents clearly stated the importance of mining activities as a factor in the degradation of soil, air, and water resources.

Moreover, these studies defined critical environmental zones (CEZs), that is, areas in which major processes of environmental degradation have become or are on the brink of becoming irreversible. The 1992 national report, which updated the CEZs defined in the previous ONERN study, identified 16 CEZs. Of these, eight have mining activities as the main degrading factor and two (Cerro de Pasco – La Oroya and Tambo–Ilo–Locumba) have mining and metallurgical activities as practically the only economic activities causing environmental degradation. No other production industry shows such a widespread incidence in the CEZs.

The CEZs that have mining activities as the main originating factor (CEZMs) are presented in Table 1. From this information, it may be concluded that the most commonly affected features are watercourses. This reflects the findings of two pieces of research: a study of the rivers of the Pacific and Atlantic basins undertaken by ONERN to produce a national diagnosis of water quality for a national plan for the use of the country's hydraulic resources: and a study integrating a group of studies on pollution and preservation of important river basins, such as Moche, Mantaro, Rimac, Santa, and Hualgayoc–Maygasbamba–Llaucano, commissioned by DIGESA. This research was produced in the 1980s.

The most commonly quoted sources of natural watercourse contamination are flotation tailings, followed by mine waters. Flotation is the standard concentration-processing method for producing base metals, presently used by the bulk of Peruvian mines. It should be recalled here that technological change has not moved to replace this process, which was introduced in the 1920s. Instead, technical change has expanded the capacity of the mines to provide important productivity increases and so maintain their competitiveness but prevented the emergence of economically viable technological alternatives. Technological changes have not occurred in the processing of sulfides, which represent the bulk of the available base-metals resources in Peru; nor has there been any change in either the generation of flotation tailings or their environmental implications (Núñez 1991).

Table 1. Critical environmental zones with a major mining or metallurgical component.

Zone	Region [a]	Main sources of pollution (polluted resource)
Chimbote–Santa	C–H	• Steelmaking (air) • Mining (rivers) • Other
Chillón–Rímac–Lurín [b]	C–H	• Mining tailings — Pb, Cd (rivers) • Manufacturing
Tambo–Ilo–Locumba	C–H	• Mining tailings (rivers) • Copper smelters — SO_2 emissions (air)
Trujillo–Moche	C–H	• Mining tailings (rivers and sea)
Cajamarca	H	• Mining tailings (rivers)
Cerro de Pasço – La Oroya	H	• Mining tailings (rivers and lakes) • La Oroya metals complex — SO_2 emissions and residual gases (air) • (Degradation of flora and fauna)
Huancavelica–Ayacucho	H	• Mining tailings (rivers)
Puno	H	• Mining tailings (rivers)
Madre de Dios	A	• Gold production (rivers)

Source: ONERN (1986, 1991), UNCED (1992), and interviews at Oficina Nacional de Evaluación de Recursos Naturales (ONERN, national office for the evaluation of natural resources).
[a] C, coast; H, highlands; A, Amazonia.
[b] This has not been included by ONERN or the United Nations Conference on Environment and Development as having a major mining component, but Direccion General de Salud Ambiental (general directorate of environmental health) (DIGESA n.d.) has explicitly referred to mining as a main factor degrading the water resources of the Rimac river.

Two of the three CEZMs have atmospheric pollution generated by extractive metallurgical activities: Cerro de Pasço – La Oroya, where Centromín Perú has a 70-year-old metallurgical centre, and Tambo–Ilo–Locumba, where the Southern Peru Copper Company (SPCC) has a copper smelter that has produced since 1960. The third CEZM with a major pollution problem is Chimbote–Santa; however, its pollution originates in a downstream metallurgical activity (steelmaking) and fishmeal production, rather than in an extractive metallurgical activity.

These factors fully cohere with the results of a study commissioned by DIGESA, "Diagnosis of Air Pollution Sources in Peru: Bases for a National Air Quality Surveillance" (Olórtegui 1989). This study aimed "to identify and locate the most important sources of atmospheric pollution in the country" (Olórtegui 1989, p. 1), focusing on "the main cities and industrial centres causing evident levels of atmospheric pollution" (Olórtegui 1989, p. 18). Under the previously mentioned conditions, Olórtegui chose seven geographical zones, five of which overlap with the CEZs defined by ONERN. Furthermore, of these five, four — Lima, Ilo, Chimbote, and La Oroya — coincide with CEZMs pointed out above.

More specifically, this study singled out the cases of Minero Peru's Cajamarquilla zinc refinery, located only 24 km from Lima; SPCC's Ilo copper smelter; Sider Peru's steelmaking plant; and Centromín's La Oroya metallurgical complex.

Air pollution from extractive metallurgy in Peru is particularly a consequence of base-metals production. Among toxic gases that may affect air quality, SO_2 is the most common and problematic. These emissions are the result of the necessary elimination of sulfur, the bulk of which is from sulfides from minerals and concentrates used by pyrometallurgical methods for processing base metals.

However, technical changes in pyrometallurgical methods have advanced both in-plant and end-of-line solutions for environmental control. For example, in copper smelting, which is of major importance to Peru, the old reverberatory furnaces can be replaced by new alternatives that integrate various extractive metallurgical processes and also save energy. Such changes increase productivity. They also reduce SO_2 emissions sufficiently such that the off-gases are amenable to the environment. The alternative is to modify reverberatory technology; this can be done with Corporación Nacional del Cobre's (CODELCO, national copper corporation) El Teniente modified converter, as a midway solution for the environmental control of SO_2 emissions. The end-of-line solutions control SO_2 emissions by converting, according to the particular parameters, sulfuric acid, liquid SO_2, and elemental sulfur (Núñez 1991).

Mining is a crucial activity in the Peruvian economy but has shared an important responsibility for environmental degradation of the countryside in this century. A natural consequence of all this is the need to harmonize the pursuit of competitiveness with an adequate environmental-protection policy. The design of such a policy will surely have to take account of the fact that environmental problems have accumulated and become more complex in the long history of Peruvian mining.

Structure and prospects of the mining industry

In this section, we present an outline of the mining-industry structure and note in more precise terms the particular type of firms behind the aggregate figures for production and environmental impacts. This will help to highlight their respective investment perspectives and to expose the external influences that influence the firms' environmental behaviour. Furthermore, it will give a more satisfactory context for the sample of firms chosen for the case studies and show that they typify Peruvian mining.

Traditionally, the Peruvian legal system has differentiated firms into three groups according to size of operation (amount processed by their concentration plants):

- Small-scale mining, less than 350 t/d;

- Medium-scale mining, 350–5 000 t/d; and

- Large-scale mining, more than 5 000 t/d.

However, this classification cannot show the actual wide distribution in the scale of firms. In fact, there is a large gap between large-scale mining and the rest of the sector. This can be demonstrated by the fact that the largest medium-scale firms may be treating between 3 000 and 3 500 t/d, whereas SPCC's two concentration plants are treating around 100 000 t/d of minerals.

At the other side of the spectrum (in a less defined way), the small-scale mining group also shows an important diversity of firms: from some that are significantly well organized and equipped, with relatively modern technology, to the larger groups of firms that work with artisanal techniques and marginalized profitability and that also enter and leave the market with the rapid rise or fall of metal prices. It also should be mentioned that small-scale mining includes the informal producers that exploit the gold placer deposits of the Amazonian region. Although this group has not been of particular importance in Peruvian mining, its numbers have been rapidly increasing since the early 1980s.

Large-scale mining controls 95% of copper, 100% of iron, and 40% of lead and zinc production. It is also important to note the majority of the medium-scale mining operations (35–40 firms) are mainly domestically owned and specialize in zinc–lead–silver production. They control well more than half of the national production of these metals. Finally, just before privatization started, in 1992, the state-owned firms were responsible for 30% of copper, 40% of lead and zinc, and 100% of iron production. State-owned firms still control all metallurgical productions, except for blister produced at SPCC's Ilo copper smelter. This clearly indicates that the state has been the most important mining producer in the country and thus provides a global picture of the large production capacities at stake in the present privatization process.

Main type of operations and production processes

Large-scale mining is dominated by open-pit operations; the exception is Centromín's operations, which are practically all underground. Medium- and small-scale firms typically operate underground mines.

Most large-scale and several medium-scale operations are well mechanized; underground operations may involve trackless or conventional systems, or a combination of both. At open-pit operations, mine planning has generally assigned specific areas for dumping marginal ores, with the expectation that they will be processed in the future when it becomes economically feasible to do so. This is the case at SPCC's Toquepala and Cuajone mines, which started production in 1959 and 1976, respectively, and have accumulated marginal minerals that will be leached and processed by solvent extraction – electrowinning (SX–EW) as part of its present 5-year 300 million United States dollar (USD) investment program.

Mineral concentration of base metals is generally performed by flotation. This involves crushing, grinding, and flotation of minerals. Technical change in this process in the past 20 years has been summarized in the following terms:

- A sharp rise of the conventional flotation-cell size (from 100–350 ft^3 [1 ft^3 = 0.028 m^3] in the 1970s to 3 500 ft^3 in 1991);

- The introduction of huge column flotation cells in the late 1980s (by 1991, these were already working at SPCC's and San Ignacio de Morococha's [SIMSA's] operations);

- Introduction of more efficient and less environmentally hazardous reagents; and

- Introduction in the early 1980s and the subsequent diffusion (Brewis 1991; Hall 1991; Núñez 1991) of automated control systems on the basis of more simple flow sheets allowed by the much larger cells (Núñez 1991).

In general, Peruvian concentration plants, at least in large-scale mining and the larger medium-scale firms, are moving to simplify their flow sheets and increase their productivity by introducing larger flotation cells and automated production-control systems. SPCC and Compañía Minera Milpo, an important medium-scale firm, typify this finding. Unfortunately, no information has been gathered about the substitution of traditional reagents for less-polluting alternatives.

As previously shown, flotation tailings are the main environmental hazard to water resources. Effective control most frequently requires the construction of special technologically designed ponds for their adequate disposal and treatment. The volume of materials is huge in a large mining operation, and the required investment may be considerable, as the case studies show.

Extractive metallurgy in Peru, like most of the world's base-metals production, follows the pyrometallurgical route. The process for eliminating sulfur from the sulfides, which constitute the bulk of the minerals and concentrates, generates toxic gases, among which SO_2 is the most common environmental hazard.

As mentioned earlier, the problem of controlling SO_2 emissions, especially from copper production, is related to the reverberatory-furnace technology, which generates this gas in concentrations of less than 4%; this is also why it is uneconomical to use the established control alternatives — neutralization through conversion to sulfuric acid, elemental sulfur, or liquid SO_2.

In the past 2 decades, several new alternatives to the reverberatory furnaces have entered industrial production. These integrate two or more extractive pyrometallurgical processes in one furnace (offering substantial energy savings) and increase the concentration of SO_2 in the off-gases, which makes neutralization economically feasible and more environmentally acceptable.

A midway solution developed by CODELCO, the El Teniente converter, is of particular relevance to developing countries with long mining histories, such as Peru. This solution involves partial changes in the reverberatory-furnace technology that increase productivity by increasing the actual capacity of the furnaces. This technology makes important energy savings and produces off-gases with an SO_2 concentration high enough to make its conversion to sulfuric acid economically feasible. The investments involved in El Teniente converter are significantly lower than those required by the new furnaces, although it seems to provide lower internal rates of return.[2] The lower investment requirements of the El Teniente converter, coupled with the difficulty most developing-country producers have in getting external financing, especially after the foreign-debt crisis, led Núñez (1991) to presume that this was a more likely choice in these countries. This has been confirmed by the information gathered through fieldwork in Peru, as will be explained in the case-studies section.

The hydrometallurgical route for base-metals production is also used in Peru, such as at Cerro Verde copper mine, developed by Minero Peru in the mid-1970s and just acquired by Cyprus Mines in the current process of privatization. If not appropriately controlled, this type of process presents environmental hazards from leakages of toxic solutions, particularly to water resources. However, it may be noted that there have been no environmental complaints of this sort against Cerro Verde since it began operation in 1977, but numerous complaints have been registered against La Oroya and Ilo copper smelters since they started production in 1922 and 1960, respectively.

---

[2] This may be concluded from a financial evaluation of smelter alternatives for Chino modernization, discussed in OTA (1988). In this case, three options were considered: installing an Inco flash furnace; retrofitting the existing reverberatory furnace; or shutting down the plant.

Lastly, gold production has been gaining in importance since the early 1980s and includes producers of gold either as a by-product or as a main product. Centromín is the main producer of gold as a by-product; gold as a main product is produced by both formal firms and informal producers (Webb and Fernández 1991, 1992, 1993). It seems reasonable to assume that the informal producers, working mainly but not exclusively in alluvial deposits of the Amazonian region, might have grown in number concurrently with this new rise of gold, because of the profound economic depression in Peru. However, research has provided no definite information on this. If this is confirmed, it also seems probable that the environmental degradation distinctive to informal production has proportionally increased, because no attempt has been made to control these producers. in fact, as early as 1986, ONERN had defined a large area in the southeast of the country, in Madre de Dios, as a CEZ (ONERN 1986).

However, the environmental implications of informal gold production are more localized and far smaller than those of base-metals production. Therefore, our research excludes informal gold mining but includes Yanacocha, a formal gold project that has been the most important gold development in Peruvian history. The environmental implications of this project, which uses hydrometallurgy, are discussed later in this chapter. This is the only case we know of in which a firm (Newmont Mining, Denver, CO) has explicitly stated that it will not only abide by Peruvian environmental regulations but, in accordance with its own environmental code of conduct, also abide by the much stricter rules of the United States.

Future perspectives of the mining sector

From the information gathered for this study, we can generally conclude that a new wave of mining investment, not witnessed since the 1950s, has started in Peru. This is led by foreign firms of diverse origin, including firms based in newly industrialized economies (NIEs). The larger domestically owned operations of the medium-scale group are also participating in this, particularly in association with foreign partners, for example, Compañía de Minas Buenaventura S.A.

The emergence of new foreign capital is mainly associated with the current process of privatization. The government requires investment commitments for the next 3–5 years from foreign bidders to modernize and expand production in the units involved. This investment requirement is not restricted to mining operations but also includes the privatization of mining deposits, as in the case of Quellaveco, a large copper deposit.

By the same token, state-owned firms have practically ceased their development plans, turning their efforts to rationalization of production. The aim of this is to regain profitability by making more efficient use of labour, materials, and installed capacity, without significant expenditures, thereby attracting private bidders. Although these firms have had no relevant development projects in the past 3 years (with the exception of Centromín's new oxygen plant), their managements were successful in this process of rationalization, as shown by Minero Peru's and Centromín's financial results for 1993 (Centromín n.d.).

Nevertheless, the dynamism observed in the mining industry in recent years cannot be exclusively accredited to privatization but also relies on a macroeconomic policy developed to promote private investments. This policy has included the elimination of restrictions on foreign exchange and major changes in labour regulations. As will be discussed later, new specific sectoral laws have also reinforced the incentives for mining investment. It is important to note, however, that these same laws have, in some instances, relaxed the environmental restrictions on the exploitation of natural resources.

The projects directly related to privatization are the following:

- *Refurbishing Marcona iron mine to recover its 10×10^6 long tons/year nominal production capacity (1 long ton = 1.016 t)* — This deteriorated over the years, reaching its lowest point in 1992, when it produced only 2.7×10^6 long tons. The mine, including its processing and auxiliary facilities, was the sole property of Hierro Peru until 1992, when it was acquired by Shougang Corporation (Beijing, China) for 120 million USD. Shougang took responsibility for the firm's debts (42 million USD) and made a commitment to invest 150 million USD in 1993–95 to recover the original production capacity. It is possible that the mine's capacity will be expanded to $15–20 \times 10^6$ t/year, and there may be a steelmaking plant constructed near the present processing facilities (Kisic 1993; Orihuela 1993).

- *Expansion of the Cerro Verde facilities from its present capacity of 36 000 lb/year to 100 000 lb/year (1 lb = 0.45 kg)* — Previously owned and developed by Minero Peru, this mine was bought by Cyprus Mines in October 1993, for 35 million USD. Cyprus Mines planned to invest 485.3 million USD in 1994–98. Buenaventura also has a 10% participation option for its development. At the time of writing, the technological choice for this project had not yet been defined (Gestión 1993).

- *Development of Quellaveco deposit* — This deposit will be developed by Mantos Blancos, Santiago de Chile, a subsidiary of Anglo American

Corporation (which consists of Anglo, De Beers, MINORCO, and Anglo American Gold Investment), provided that the results of a 2-year feasibility study, now under way, are favourable. The deposit is expected to produce 100 000 t/year of cathodic copper, probably by SX–EW.

The two main new projects not originating from privatization are the Yanacocha and Iscaycruz. SPCC's 300 million USD investment program probably responds in part to the new, promotional legal framework, but the case-study analysis indicates that it also responds to other factors, including the environmental impacts of its operations.

Of the more relevant projects not connected to privatization, the following are notable:

- *SPCC's 300 million USD investment program* — This involves both expansion and environmental projects. Expansion includes the construction of two leaching plants, which will use SX–EW processing on the Cuajone and Toquepala dumps of marginal material. This will allow SPCC to expand production by 9%. The environmental component, comprising about one-third of the total investment, includes construction of a sulfuric acid plant for partial control of the SO_2 emissions from the Ilo copper smelter and a system to control the concentration tailings. SPCC has not yet decided whether the 30×10^6 t/year tailings of Cuajone and Toquepala will be deposited on the mainland or under the sea. In any case, SPCC's program represents the largest environmental investment yet undertaken in Peru.

- *Yanacocha gold project* — When fully developed in 1995, the Yanacocha gold project will produce 500 000 oz (1 oz = 28.35 g) of gold per year. This is about 15 t/year, or 1.5 times Peru's 1992 production of gold (INEI 1993). Newmont owns 38% of the shares and is associated with Buenaventura (32.3%), the Bureau de recherches géologiques et minières (a French state-owned firm; 4.7%), and the International Finance Corporation (IFC) of the World Bank (5%). An initial investment of 36.6 million USD was needed to bring production up to 250 000 oz/year. An estimated additional 14 million USD in 1995 was needed to bring production to 500 000 oz/year.

- *Iscaycruz* — This property was developed by Compañía Paraibuna di Metais (Brazil), which holds 45% of the stock, in association with the state-owned Minero Peru (25%) and Buenaventura (15%). With an investment commitment of 39.8 million USD, Iscaycruz is expected to produce around 120 000 t/year of zinc and 10 000 t/year of lead concentrates in 1995 (Centromín n.d.).

The ongoing and expected expansions of the sector stress the urgent need for an appropriate environmental regulatory system.

The legal and institutional system for environmental control and its prospects

We undertook an historical analysis of the legal and institutional system for environmental control in Peru, especially in relation to mining and metallurgical activities. Our purpose was to study the main factors influencing the emergence of this system and the actual enforcement of its regulations. This will allow us to assess the more recent developments in setting up a new regulatory system and to evaluate the extent to which the limitations set by previous regulatory frameworks may be overcome.

The historical analysis covers 1950 to the present, with reference to the main macroeconomic policies and particularly to the role of the state. We distinguished three periods:

- *1950–68* — Peru had a typical laissez-faire economy, in which the state was practically absent from direct production activities and the economy was fully open to international competition. (There were a few exceptions, such as a small steelmaking plant in northern Peru, in Chimbote. This was SOGESA, which years later became Sider Peru.)

- *1968–90* — The state was the main entrepreneur. However, opening the economy to the world dramatically changed Peru in this period, first by introducing strong foreign control on capital movements (1968–75), then by relaxing them (1976–85), and finally by reintroducing them (1985–90).

- *July 1990 to the present* — The state has retreated altogether from production activities and will go back to having as its sole role that of promoter of private capital in the framework of a fully open economy.

We studied the general dynamism of this sector and the emergence of its regulatory system, particularly environmental controls. Our purpose was to determine whether there was any connection between distinct macroeconomic policies and the emergence of the environmental regulatory system and the actual enforcement of its regulations.

The regulations have emerged to a large extent independently of macroeconomic policies. The regulations dealing with the environment in the workplace and outside the fences of production units have been numerous (Andaluz and Valdez 1987), but their actual enforcement has been weak, to say the least. However, historical analysis showed that public opinion, availability of resources, and international technical and financial cooperation can play an important role in improving the system of environmental regulation.

The *Environmental and Natural Resources Code*, published in 1990, represented a qualitative change from previous legislation because it attempted to coherently integrate the dispersed and not infrequently contradictory legislation that preceded it. Moreover, environmental legislation after the Code has reoriented the spirit of environmental regulation from being nominally punitive to being supportive of feasible and adequately determined environmental standards. This has been particularly so in the case of legislation for the mining sector issued as the Regulation of Title XV (that is, Environment) of the *Unifying Text of the General Mining Law* (D.S. 016-93-EM), of April 1993.

This following paragraphs support the above summary.

From 1950 to the present, new environmental regulations and institutions have appeared independently of the specific macroeconomic policies.

In 1950–68, the most important progress was made in environmental conditions in the workplace. The Instituto Nacional de Salud Ocupacional (INSO, national institute of occupational health), which was responsible for this performance, is probably the most successful example of an environmental institution in the country. INSO was set up in 1947 through a cooperation program between the Peruvian and the US governments. The initial purpose was to reduce the high incidence of silicosis and other occupational diseases that had been affecting Peruvian mining workers for some time. Public awareness of these problems probably influenced the government's decision to participate in this program. Vizcarra (1982, pp. 219–220) presented a detailed account of this institution, indicating that "it was granted rents through an ad hoc law (No. 10833) that amounted to 1.8% of the mining wage bill. Although [INSO] was ascribed to the Ministry of Health it had the capacity to act with autonomy and at the national level."

The program's direction was initially controlled by foreign professionals and advisers. Later on, these were replaced by national professionals who had attended postgraduate courses in specialized centres abroad as part of the program. Particularly relevant here is that INSO produced technical reports and studies on occupational health at 120 mines in Peru. INSO then widened its scope, undertaking studies in manufacturing and agricultural activities. The success of INSO also spread to other Latin American countries, as it became, on the basis of its national achievements, a recognized training centre.

Although Vizcarra (1982) emphasized that the studies developed by INSO included specific recommendations for confronting the problems detected, he provided no information about the ways firms actually responded to these recommendations. In any case, the successful trajectory of INSO came to a halt in the 1970s, when its integration with other health institutes eliminated its specific rents and the important autonomy it held. This case shows that international cooperation, availability of resources, and adequate autonomy can play a very important role in the performance of environmental institutions.

History shows that the enforcement of environmental legislation in Peru has clearly been weak. A study undertaken by the National Council for the Protection of the Environment for Health (CONAPMAS), which became the nstituto Nacional de Protección del Medio Ambiente (national institute for environmental protection), grouped together the accusations (of contamination) filed against mining firms before DIGESA at the Ministry of Health and at the Office of Environmental Affairs by the Ministerio de Energía y Minas (MEM, ministry of energy and mines) in 1970–87 (Conapmas 1988). The study found that only in 11 of 64 cases was there a judgment (6 in favour of the claimants and 5 against). The other 53 cases received none. In other words, the regulations were not of much use in more than 80% of the cases. Similarly, another study on the application of forest regulations revealed that only 15% of cases were subject to judgments under the relevant regulations (Andaluz and Valdez 1987). Among the explanations for the weak enforcement of the environmental legislation are the following:

- Not infrequently the scope and jurisdiction of the regulations have been ill defined.

- In some cases, the regulations contain evident mistakes; some are just simple typing errors. In most cases, however, these could have been easily corrected, but even in important instances they have not been. For example, the water-quality standards for the country, which are contained in the Regulation of the *General Law of Waters* (D.S. 007-83-SA), are 1 000 times stricter concerning sulfate content than those of the Environmental Protection Agency in the United States. This has given rise to distrust about the technical and scientific backing of the whole regulation, weakening the basis for its enforcement.

• Although the number of environmental regulations is high, there are clearly important gaps. When the actual terms with which firms must comply are not fully and clearly specified, it is difficult to enforce the standards. This is the case with environmental standards. Existing standards in Peruvian legislation pertain to the environment in the workplace and to water quality. No standards have yet been defined for effluent, air quality, or emissions. Since 1984 different drafts of regulations on air standards have been proposed to the highest levels of government and to Parliament, but nothing in this field has been finally approved.

• Environmental institutions lack the resources to do their job. In the case of INSO, the importance of this has already been emphasized. Also, DIGESA, which commissioned and produced valuable studies in the mid-1980s, showed a dynamism that decreased rapidly toward the end of the decade (as happened in most of the public sector), resulting in a reduction of its budget.

• The legislation and the environmental institutions have a sectoral focus. This is because a ministry covers specific types of activities, without a suprasectoral level of coordination to ease the application of the regulations.

• The pre-1990 legislation emphasized only penalties for firms that failed to abide by it, without offering guidance to those that wanted to comply with it. This led to poor application of regulations and even to corruption. However, the new regulatory framework, following the introduction of the Code, has shown qualitative changes, particularly in mining and petroleum activities. Thus, the Programa de Adecuación y Manejo Ambiental (PAMA, environmental-management and adequation program) was introduced. Its purpose was to open up ways for production units to achieve the appropriate environmental standards.

The present environmental system as it applies to mining activities

At the time the *Environmental and Natural Resources Code* was published (September 1990), the main environmental regulations related to mining-sector activities were

- The Regulation on Mining Safety and Welfare (D.S. 034-73-EM/DGM), published in 1973, which stipulated environmental standards for mines and metallurgical plants;

- Water-quality standards, which were set by the Regulation of the *General Law of Water* (D.S. 261-69-AP) in 1969, as well as its later modifications in 1983 (D.S. 007-83-SA);

- Other regulations, such as the *Forest and Wild Fauna Law* (D.L. 21147) of 1975, which strongly limited the exploitation of natural resources in zones declared as conservation areas (that is, national parks, national reserves, national sanctuaries, and historic sanctuaries), and the *Sanitary Code* of 1969.

The mining sector was regulated by the *General Mining Law* (D.L. No. 109) of 1981, within the framework of the 1979 National Constitution, which was the first to include articles specifically addressing the environment. One of the 12 articles (art. 123) expressed the following view:

> Everyone has the right to inhabit a healthy, ecologically balanced environment that is adequate for the development of life and the preservation of landscapes and nature. Everybody has the duty to conserve that environment. It is an obligation of the state to prevent and control environmental contamination.

From an analysis of the Code's content, it seems clear that one of its aims is to overcome the dispersed and sector-specific character of previous environmental legislation, which had been so poorly enforced. Also, the chapter XXII of the Code created the National Environmental System, which, according to art. 128, was to be

> made up by all public institutions dedicated to research, evaluation, monitoring, and control of natural resources and the environment and by the departments and offices of the different ministries at the national, regional, and local levels that perform similar roles. By a Supreme Decree the government shall determine the co-ordinator of the System.

The Code, however, contained shortcomings. For instance, production activities directly affected by the Code argued that it lacked the necessary technical backing to be valid and therefore that it had to undergo major changes before it could be enforced. The outcome of this was the annulment of major sections of the Code, which was performed indirectly by the promulgation of laws specifically developed to promote private investment. These laws were the *Frame-work Law for the Growth of Private Investment* (D.L. 757) and the *Law of Investment Promotion in the Mining Sector* (D.L. 708), both published in November 1991. The sections of the Code superseded in this way included the one defining the National Environmental System (chap. XXII) and one defining the penalties for violations (chap. XXI).

No substitute legislation has been issued for these sections. Meanwhile, the applicability of the Code has been substantially undermined. Moreover, the intended suprasectoral nature of the Code has been further undermined by the emergence of new sectoral legislation for mining and petroleum activities that names MEM as the authority to issue the maximum permissible levels of environmental control. This is expressed in a complementary ruling of the Regulation of Title XV of the *Unifying Text of the General Mining Law* (D.S. 016-93-EM), published in April 1993. This regulation constitutes the most, and perhaps the only, specific environmental legislation for mining activities. It specifically addresses environmental controls in their respective areas of influence.

In October 1992, the Regulation of Mining Safety and Hygiene (D.S. 023-92-EM) was published, superseding the Regulation on Mining Safety and Welfare of 1973. Both explain, among other things, the minimum environmental standards for workplaces at mines and metallurgical plants. The Regulation of Title XV was introduced by the PAMA for mining activities and placed its emphasis on providing ways to make the existing activities comply with environmental standards, rather than merely penalizing offenders, as was the case with previous legislation.

The PAMA's concept is not included in the Code, but it was included in the official Proposal for the Debate on the Regulation of the *Environmental and Natural Resources Code* published in 1991. This may give some support to the opinion that approach should have been to iron out the shortcomings of the Code through this regulation, rather than annulling entire sections of the law. It was this latter approach that subsequently limited the Code's applicability. It should be added, however, that this regulation has not yet been issued.

In 1991, the administration redefined the state's role in production, that is, that the state should not be present at all in direct production but should return to its sole function as the promoter of private capital. This role was specifically expressed in the promotional laws. In addition, the purposeful introduction of new environmental legislation led to the rewrite of the 1981 *General Mining Law*. This new version was issued as the *Unifying Text of the General Mining Law* (D.S. 014-92-EM) in June 1992.

The Regulation of Title XV and its modification in December 1993 (D.S. 059-93-EM) represent the essence of the present environmental legislation for mining operations:

- All existing mining operations are required to present an annual environmental-impact declaration (EID).

- The operations must monitor their activities for 1 year, producing a preliminary environmental assessment. Its approval or rejection by MEM may take up to 3 months. The ad hoc formats to elaborate this assessment will be available only as of March 1994.

- After approval of the preliminary assessment, the operator has 1 year to produce a PAMA. Its approval may take up to 6 months. For mining and mineral-processing activities, the approval of the PAMA may be good for up to 5 years; for other downstream metallurgical operations, up to 7 years.

- The annual investment involved in the PAMA must be equivalent to at least 1% of total sales.

- New operations should present an environmental-impact assessment (EIA), which has to be approved by MEM.

From this timetable for environmental compliance, it may be gathered that no actual investment in compliance with environmental legislation (that is, the actual application of the PAMA) could be expected before 1997.

Lastly, it should be stressed that the PAMA and the EIA have to be under-taken by third-party firms registered for these purposes with the Direccion General de Asuntos Ambeintal (DGAA, general directorate for environmental affairs) of MEM. Alternatively, the preliminary environmental assessment and EID may be performed by the firm itself, but they must be assessed by an environmental auditing firm registered with the Direccion de Fiscalización Minera (DFM, directorate of mining control), also at MEM. Thus, although the new environmental regulations for mining activities became wider in scope, MEM relinquished its direct control, as this responsibility has been transferred to private firms or public institutions (that is, universities). This fully coheres with the governmental directive that the state's influence should be sharply reduced. However, it seems certain that a minimum controlling capacity will be held at the ministry, from which it will be able to undertake random inspections of the controllers. From information gathered from the DGAA and DFM in June 1993, we got the impression that rather than keeping a minimum capacity to carry out random compliance inspections, these offices would assign that task to the third-party firms registered at both offices.

DGAA, a branch of MEM created in 1981, produced a high number of technical environmental reports on mining units in the 1980s. International cooperation, particularly that of the Japan International Cooperation Agency (JICA) (Cacho, personal communication, 1991[3]), was crucial for the development of this procedure. However, DFM has transferred these types of activities to firms or high-level academic institutions, keeping only the role of overseeing the work.

DIGESA, at the Ministry of Health, is in charge of controlling environmental issues (that is, water, air, and soil) with respect to human life. The research it commissioned in the past has already been cited. Among DIGESA's duties is monitoring water-quality standards to protect human health. In this respect, it can intervene if a mining-pollution incident takes place. However, the 1993 Regulation of Title XV, granting MEM all responsibility for approval of environmental standards for mining activities, seems to have created a conflict in this matter.

ONERN used to be a decentralized public organization, dependent on the National Planning Institute. However, when the latter was eliminated and the organizational charts of the ministries were restructured, ONERN amalgamated with other departments of the Ministry of Agriculture, under the umbrella of the Instituto Nacional de Recursos Naturales (INRENA, national institute of natural resources). INRENA is in charge of the "management and the rational and integral use of the renewable national resources and the ecological environment to achieve a sustainable development" (D.S. 055-92-AG, art. 4). Furthermore because mining activities, like almost any other production activity under certain circumstances, may represent a threat to the conservation or preservation of renewable natural resources, this institution has the obligation to regulate these activities.

Domestic technological capabilities in environmental control of mining activities

Peru is a country with a long mining tradition, during which it has kept a significant presence in the world mining industry. The importance of mining in the national economy and its connection with other sectors show that it has created important domestic demand for its material inputs and its qualified personnel.

---

[3] N. Cacho, director of DGAA at MEM, Lima, Peru, personal communication, 10 January 1991.

The response of the production system to this demand is reflected, for example, in the number of universities in Peru that offer courses in mining and metallurgic engineering, as well as geology. In addition, the country has several related organizations and professional associations, and these groups publish specialized journals and organize seminars and conferences.

This dimension of the sector and its long history in production have brought about the development of relevant technological capabilities, not only in mining but also in the production of certain capital goods and inputs, as well as in the provision of technical services for the industry. This is shown by the significant number of local equipment suppliers and, to a lesser extent, of mining consultant firms. For example, a catalogue of equipment suppliers for mining operations and concentration plants, produced by the Board of the Cartagena Agreement (the Andean Integration Organization) in the late 1980s, lists 20 Peruvian firms (Junac n.d.). On the other hand, in 1983, eight or more domestic consulting firms were officially registered with the Corporación Financiera de Desarrollo S.A. (COFIDE, development finance corporation) as being eligible for state contracts in the areas of prospecting, mining, and metallurgy. More consulting firms were registered at the time in the areas of environmental sanitation and water and waste treatment. Also, some domestic and foreign firms appeared as specialists in geology, seismology, hydrology, and meteorology.

In nominal terms, this may lead us to presume that the domestic technological capabilities relevant to environmental management in the mining sector might be significant and should be considered in the design of any environmental strategy. With this premise, we examined a small sample of consulting firms and two of the most important university centres. We gathered information through direct interviews with top representatives of these firms and institutions.

The sample of consulting firms

In accordance with the terms of the new environmental regulations for mining activities, MEM maintains two registers of firms: one for firms officially eligible to conduct EIAs and PAMAs at the DGAA; and one for firms similarly eligible to conduct mining auditing tasks for the EVAPs and EIDs at the DFM.

In May 1993, the initial idea was to compare in a complementary way the information available in COFIDE's register (which had existed for 10 years) with the information in the registers set up by MEM in the first quarter of 1993. However, it turned out that no updated list of processed firms (that is, classified by type of specialization) was available at that time at CONASUCO (the office at COFIDE in charge of the register) (Ramírez, personal communication, 1993[4]). On a different account, the register at DFM had only a couple of firms registered and provided an extension to new applicants.

Eventually, a sample of 4 firms was chosen from the list of 25 registered at the DGAA at MEM in May 1993 (Lanza, personal communication, 1993[5]). Two were chosen because they or their main representatives already appeared as specialists in COFIDE's 1983 register, namely, Buenaventura Ingenieros S.A. (BISA) and Aqua Plan Ingenieros. The other two were Laboratorio Geotécnica S.A. (LAGESA), with more than 25 years' experience in consulting work (although it is new to environmental work) and Ecología y Tecnología Ambiental, a new consulting firm.

It should be emphasized that BISA, a firm belong to the Compañía de Minas Buenaventura group, and Instituto Geológico, Minero y Metalurgico (INGEMMET, geological, mining, and metallurgical institute), a government entity, were the only two registered at the DGAA that had also appeared in COFIDE's register as specializing in mining and metallurgical activities by 1983.

This inquiry was aimed to provide some very preliminary information about the role domestic technological capabilities could play in environmental-management tasks stipulated by the new legislation. From the analysis of the sample of the consulting firms, the following may be concluded:

- No relevant experience on environmental control for mining activities seemed to be available yet. BISA was the only consultancy with experience in mining activities and was the most important domestic firm in this area: it has had several significant international projects (participating in the development of mining projects in Venezuela, Argentina, Colombia, and Ecuador).

- The experience in environmental engineering was limited to water control. No previous work on atmospheric control was reported by these firms. It should be noted that this specialization was not even considered in COFIDE's 1983 register.

- The consulting firms seemed to have very limited laboratory facilities of their own, relying instead on those available at universities or specialized public institutions. This is apparently not an uncommon

---

[4] Lic. Javier Ramírez, head of CONASUCO, interview, 10 May 1993.

[5] Ing. Jorge Lanza, general director of DGAA at MEM, interview, 10 May 1993.

procedure in consulting work: BISA, for example, for its work on mining activities, frequently used the laboratories at the Universidad Nacional de Ingeniería (UNI, national university of engineering) and at
INGEMMET. LAGESA also stated that for its environmental work, it
used the laboratories at SEDAPAL, Lima's potable-water firm.

• All four firms did other kinds of environmental-control work of interest,
apart from mining. For example, BISA was registered to develop EIAs
in energy and industrial activities; LAGESA expressed similar capabilities for the electricity-generating industry; and Aqua Plan specialized in
environmental sanitation and water and waste treatment generally, a
specialization it had maintained for more than 2 decades.

• There seemed to be some consensus among the firms that the
qualifications of local professional groups for mining and for environmental activities related to water were quite important. The point is
that those capabilities were only then being pulled together in response
to the new regulatory framework for the mining sector. BISA assured
us that, because of its professions qualifications and experience, the
firm was not only locally but also internationally competitive. BISA
indicated that when a very specialized technical capability was needed
that was unavailable locally, the firm made the necessary contacts to get
it from abroad. BISA was going to approach environmental protection
in a similar way (Benavides de la Quintana, personal communication,
1990; Sánchez Saavedra, personal communication, 1993[6]). Similarly,
Fernando Chuy Chang, managing director of Aqua Plan, with more than
25 years of professional experience in consulting on environmental engineering, confirmed that the local professional groups in this area were
quite good for these types of studies. As in the case of BISA, when it
was necessary to complement the local team with foreign expertise,
Chang had generally been able to find it. However, in Chang's opinion,
there was an important gap in qualified mid-level technicians, particularly those who work in the specialized laboratories. For environmental
auditing, it was necessary to have not only adequate infrastructure, but
also specialized personnel for adequate operation and maintenance.
Chang mentioned instances of studies having been spoiled because of

---

[6] Ing. Alberto Benavides de la Quintana, president, BISA, interview, 27 December 1990; Ing.
Jaime Sánchez Saavedra, manager, BISA, interview, 18 May 1993.

inadequate calibration of instruments (Chang, personal communication, 1993[7]).

These points give the impression that certain relevant domestic capabilities outside the mining firms were needed for the environmental-management tasks (EIAs, EIDs, etc.) stipulated by the new regulatory system, but the external component needed to adequately perform these tasks is likely to be important, particularly to atmospheric-pollution control. The participation of domestic techno-logical capabilities, as in the past with the supply of mining equipment and con-sulting services, will probably concentrate on the market provided by medium- and small-scale mining.

University centres

We also undertook a preliminary examination of two universities to determine whether their present human resources and laboratory infrastructure had the capacities to confront the more relevant environmental problems and to take responsibility for carrying out role particular tasks defined by the new regulations. Our research also aimed to provide a limited account of the slower development of technological capabilities through the formal system of higher education.

To choose the two universities and programs to be analyzed, we set out the following criteria:

- The programs had to directly relate to environmental control in mining and metallurgical activities; and

- The sample universities had to be from the vicinity of Lima, which had the most important university centres.

With this perspective, we approached the Statistics Office of the National Assembly of University Vice-Chancellors (ANR) in May 1993. A catalogue of the programs available in the national university system in the country had just been completed. From the analysis of this information, we chose two universities: UNI and the Pontificia Universidad Católica del Perú (PUCP, Catholic University of Peru). UNI had programs on geology, industrial hygiene and safety, sanitary engineering, and mining and metallurgical engineering, as well as a master's program on sanitary engineering. PUCP had a well-equipped mining engineering program and, in 1992, had entered a Technical Co-operation Programme with

---

[7] Fernando Chuy Chang, managing director, Aqua Plan Ingenieros, interview, 30 June 1993.

Cardiff University, supported by the British Council, to develop capabilities in environmental management of mining operations. PUCP decided to provide its excellent institutional support for our research precisely because of its particular interest in the topic of mining and the environment. Furthermore, in 1993, PUCP decided to set up the Institute of Environmental Studies (IEA) to pull together the work of different departments of the university and strengthen its future work on the environment with an adequate multidisciplinary approach.

It should also be pointed out that, if our intention had been to provide more than a preliminary study, our sample would have included other universities, such as the Universidad Nacional Mayor de San Marcos and Universidad Nacional Agraria, which, according to the information provided by ANR, also have particularly relevant programs.

At UNI, the information gathering was limited to the Faculty of Mining, Metallurgy and Geological Engineering (FIMMG) and the Faculty of Environmental Engineering (FIA). It is interesting to note that UNI began its activities with a School of Mines and Civil Works Construction well over a century ago, in 1876. studies related to the environment started in 1937, with the Faculty of Sanitary Engineering, which in 1984 became FIA. In 1973, this faculty established an undergraduate program in engineering of industrial hygiene and safety.

Both FIMMG and FIA have master's programs: FIMMG has one in mining engineering; FIA has one in water treatment and waste reuse and another (jointly with Universidad Mayor de San Marcos) in occupational health and hygiene.

PUCP was founded in 1918, and its Engineering Mining Section started its activities in 1970. To set up of this section, PUCP counted on major support from the United Kingdom, through the University of Cardiff. In fact, 2 decades later, Cardiff renewed its support, this time to add environmentalists specializing in mining activities to the technical groups. We obtained information about the Engineering Mining Section from two important centres at PUCP, both directly related to environmental-control issues: the Centro de Investigation en Geographia Aplicad (CIGA, centre for applied geographic research) and the Laboratory of Corrosion.

An important omission from our research — the chemistry programs at both universities — should be acknowledged because chemists have important roles in environmental control. Only the very preliminary nature of this inquiry justifies this omission.

Our analysis of the information provided indicated the following:

- The curriculum for mining and metallurgical engineering included only one course on environmental issues, on safety and hygiene in mines and plants (that is, environmental conditions at the workplace). However, in a couple of exceptional instances, a course contained topics related to environmental control. For example, a course on auxiliary services included the design of tailings ponds. At both universities, interdepartmental coordination was strengthening the presence of environmental issues in the curriculum. At UNI, this was being done in FIA and the Faculty of Chemical Engineering. At PUCP, IEA was set up to connect and integrate the different groups working on environmental issues at that university.

- FIA's curriculum included the elaboration of EIAs. FIA and the FIMMG were collaborating to develop an EIA for the polymetallic Iscaycruz project, as well as for expansion of the Ishihuinca gold mine, under contract to BISA.

- FIA's postgraduate programs have been changed to address water and waste control, as well as occupational-health issues. Atmospheric-pollution control appears not to be a primary concern at this level.

- At both university centres, steps were being taken to officially participate as consults in environmental issues. To this end, UNI was establishing the coordination of FIA and UNITEC, its consulting and services arm. PUCP was planning to do the same with IEA and the Centre of Technological Transfer.

- The relationship between automation and the capacity for environmental control is well known. The mining section of PUCP was working to set up a laboratory for automation of flotation plants for academic as well as consulting purposes. Because of the main importance of flotation processes in mining production and its environmental implications, this step appears to be of major relevance.

- Remote-perception techniques have a significant capacity to monitor water and soil contamination, the direction of emissions into the atmosphere, and other processes. CIGA (at PUCP) was working with geographic information systems (Bernex and Córdova, personal communication, 1993[8]) and remote-perception equipment in several

---

[8] Nicolle Bernex and Hildebrando Córdova, CIGA, interviews, 27 August 1993.

areas of the country (such as, Tambo Grande). It is important to note that CIGA was operating with five full-time researchers and providing training programs, in addition to the support they gave to other departments at PUCP. CIGA has relevant backing from Belgium cooperation.

- The laboratories were far better equipped for doing research on water-quality control than on air-quality control.

FIA had three laboratories: one for sanitary engineering (physicochemical and bacteriological); one for sanitary machinery and equipment; and one for ergonomics, hygiene, and safety. FIA was trying to modernize these laboratories not only to support its academic studies but also to provide external services. Moreover, FIA intended to become a reference laboratory in those areas, for which it was looking for support from the Inter-American Development Bank (IDB). FIA's Sanitary Engineering Laboratory already has the capacity to determine water quality according to the parameters set up by the Regulation of the *General Law of Waters*, described above (Botto, personal communication, n.d.[9]).

UNI pointed out that it had a capacity to undertake a wide range of atmospheric studies (Sotillo, personal communication, 1993[10]). However, it was unable to do analyses using portable monitoring equipment, particularly important for work in chimneys in industrial plants. IDB was asked to assist UNI's efforts to acquire this equipment.

At PUCP, the Engineering Mining Section's laboratories were set up with the specific purpose of determining assays of minerals with economic value, and these laboratories do not have the capacity to detect traces of contaminants. However, PUCP's Laboratory of Corrosion had the capacity to perform this kind of analysis for water-quality determination (Díaz, personal communication, n.d.[11]). It was probably the best-equipped corrosion laboratory in the country. Set up in 1989, it depended on the support of the German Gesellschaft für Technische Zusammenarbeit, which grants resources for training professionals at a post-graduate level in Germany and for laboratory facilities.

For PUCB's studies of air-quality control, the Engineering Mining Section depends on equipment on loan from MEM. This equipment was donated several

---

[9] Ing. J. Ruiz Botto, dean, FIA, interview, n.d.

[10] Ing. Francisco Sotillo, UNI, 1 June 1993.

[11] Ing. Isabel Díaz, head, Laboratory of Corrosion, PUCP, interview, n.d.

years ago to MEM's DGAA by JICA. Some of this equipment was appropriate for SO_2 monitoring and had been taken to La Oroya for that purpose.

However, we received the impression that the capacity was again much weaker for air-quality than for water-quality control. It should be added that according to a provisional inventory in the country, 47 laboratories had the capacity to determine water quality; 20% of these were in Lima, and 89% were in the rest of the country.

Case studies: mining firms and the environment

A central objective of this research was to study the attitudes of mining firms toward the environment and the environmental regulatory framework. Moreover, we wished to determine the extent to which particular patterns of response can be ascribed to particular types of firms. We approached these questions through the detailed analysis of case studies. However, the scope of the fieldwork was much broader than initially planned. Thus, it practically covered all the large-scale mining firms, three of the most important medium-scale mines, and the two most important greenfield projects under development in 1993. Similarly, the sample included firms with diverse controlling interests — state, foreign, or domestic private. Also, we distinguished between foreign interests from developed countries and those from other parts of the world.

The importance of the mining sector in Peru's economy is expressed in, among other forms, the prominence of its firms compared to those of other sectors. Table 2 shows the mining firms that appeared among the 100 largest, by sales, in 1989, just a year before the state-owned sector was called into question by the 1990 administration. Thirteen mining firms are included in total. In fact, if Minpeco, the state-owned minerals-trading company, is not counted as a productive firm, then Centromín, SPCC, and Minero Peru would occupy, respectively, second, third, and sixth positions (the largest firm in 1989 was Petro Peru, the state-owned oil producer). These firms, along with the then state-owned Hierro Peru, now Shougang Hierro Peru (15th), all large mining firms, were the case studies in our research.

SIMSA, Milpo, and Buenaventura, ranking 29th, 37th, and 72nd, were included as case studies of the medium-scale firms. The relative weight of the Buenaventura group is unclear in Table 2 because it differs from the others in having more than one important mining firm. Thus, for example, the group owns Compañía de Minas Orcopampa S.A, which ranks 36th. The Buenaventura group's aggregate sales would place it at the top of the ranking of medium-scale firms.

Table 2. Peru: mining firms included in the ranking of the 100 largest firms in 1989.

General ranking	Mining ranking	ISIC	Firm	Type [a]	× 10⁹ intis Income	× 10⁹ intis Assets	× 10⁹ intis Equity
2	—	6120	Minpeco [b,c]	S–L	2 103	—	—
3	1	21, 23 29, 36	Centromín Perú S.A. [b]	S–L	2 026	2 593	935
4	2	23, 36	Southern Peru Copper Company	P–L	1 692	3 098	1 725
7	3	21, 23 36	Minero Peru [b]	S–L	860	2 555	465
15	4	23	Hierro Peru	S–L	384	—	—
21	5	23	Tintaya S.A.	S–L	344	1 460	697
29	6	23	San Ignacio de Morococha	P–M	260	258	103
36	7	23	Minera Orcopampa S.A.	P–M	202	300	46
37	8	23	Compañía Minera Milpo S.A.	P–M	201	269	90
59	9	23	Minera Atacocha	P–M	144	205	99
68	10	23	Perubar S.A.	P–M	126	138	77
72	11	23	Compañía de Minas Buenaventura S.A.	P–M	116	223	32
81	12	23	Compañía Minera Raura S.A.	P–M	100	93	32
87	13	23	Minsur S.A.	P–M	94	175	92

Source: *The Peru Report* (1990).
Note: ISIC, International Standard Industrial Classification.
[a] P, private; S, state owned; L, large-scale mining; M, medium-scale mining.
[b] Firms taken as case studies.
[c] This is the state-owned minerals-trading firm and therefore is not considered in the ranking of productive mining firms.

The sample included the Yanacocha gold and Iscaycruz polymetallic green-field projects to allow us to evaluate the responses of firms with this type of project to the new environmental regulatory framework. These cases also offered us the opportunity to examine whether the responses of firms with capital based in NIEs, such as the Brazilian Paraibuna Metais, which controls Iscaycruz, overlap or differ from that of Newmont, of the United States, which controls Yanacocha.

The respective longevity and historical ownership structure of the firms in the case studies showed that Peru had practically no state-owned production sector until the 1970s. Thus, SPCC, Cerro de Pasço, and Marcona Mining Company were already in operation when Velasco's military regime started its process of nationalization. As a result, Cerro de Pasço became Centromín Perú S.A. in 1974, and Marcona Mining Company became Hierro Peru in 1975. SPCC, on the other hand, remained untouched and signed a new agreement with that government to develop the Cuajone project, which more than doubled its previous production capacity. Hierro Peru returned to foreign control at the end of 1992, but this time its buyer was Shougang Corporation.

The state-owned sector was by no means restricted to taking over production facilities; it also decided to develop new projects. For this purpose, it set up Minero Peru and entrusted it with the state's interests in the mining sector. This firm received the mine deposits that returned to the state from private holders who failed in 1970 to present concrete plans for developing them. Minero Peru eventually developed three main projects: Cerro Verde copper mine, Ilo copper refinery, and Cajamarquilla zinc refinery. Cerro Verde was privatized at the end of 1993, but the refineries (Anon. 1994) and many mine deposits remained in Minero Peru's hands.

Firms outside the large-scale mining sector were not touched by the process of nationalization in the 1970s; in fact, new domestic firms continued to appear during that period. The most important of these is probably SIMSA, currently the largest private zinc producer in the country. This firm, Buenaventura, and Milpo, which had entered the sector 2 decades earlier, were always controlled by well-known groups of domestic investors.

Lastly, Minera Yanacocha and EME Iscaycruz, which were set up to develop greenfield projects after the 1990 introduction of the Code, are both foreign controlled, although each has an important domestic stake.

A final report of the research described, at length, all these firms and their respective units, production processes, environmental implications, and the ways they have responded to the evolving regulatory framework (Núñez-Barriga and Castañeda-Hurtado 1994). Here, the discussion will focus only on the main features that came out of the research as having some bearing on the environmental management of the differentiated type of firms (see Table 3). This comparative presentation of the results is developed below.

International pressure

International pressure (particularly in relation to financial facilities for modernization and new projects), as well as the realization by investors (international, state, and domestic private) and governments that there is little chance of this pressure being relaxed in the foreseeable future, has been crucial to the emergence of a new, much more realistic environmental regulatory framework. For the same reasons, firms have been showing a clearer stance toward compliance.

At least one important supporting factor has made local producers more sensitive to external pressures, particularly since the early 1980s. This is the difficulty firms have had getting the necessary financial resources to maintain their competitiveness in an increasingly competitive market in the context of a domestic economic policy in the second half of the 1980s that was unfavourable to the sector and of a foreign debt crisis that started in the mid-1970s. In this context, the new environmental stands of multilateral agencies, such as the World Bank, IDB, and development-aid agencies of developed countries have been granted greater attention.

It is important to note that these developments affected all types of firms. Particularly relevant has been the case of medium-scale mining firms, which not only express their influence by their relative weight in production among domestic firms but also dominate the local organizations of the mining community, whose opinions are highly respected. Domestic producers have traditionally specialized in zinc–lead–silver production, in which the weight of silver (in terms of value) has been quite important. With the depression of base-metals prices, domestic firms had to add the burden of historic lows in silver prices.

In these circumstances, investors had failed since the early 1980s to maintain the rhythm needed to face a more competitive market. From the case studies, it can be concluded that there is a need for investments to modernize production capacity and that much of this will have to come from abroad and be used for more appropriate environmental management. This has been explicitly recognized not only by the state-owned firms that receive important support for privatization from multilateral agencies but also by the medium-scale firms interviewed. Given their leading role in the mining community, their perception is likely to have a significant effect on the attitudes of smaller formal firms.

Ownership, size, and longevity of the firms; technology vintage; type of environmental solution

The information gathered indicates that environmental behaviour was unrelated to ownership structure (that is, foreign, state, or domestic private) or the size (that is, large or medium) of firms. More clearly relevant was, for example, the longevity of production capacities. For instance, of the 91 years that the former Cerro de Pasço Co. has been in operation, only for the past 20 (since 1974) has it been a state-owned firm (Centromín). Centromín inherited serious environmental problems, which had accumulated during more than 70 years of foreign ownership, but state control did not change this situation much. Only in the past couple of years, ironically, with privatization and the pressure of multilateral agencies, has the firm taken some initial steps to improve its environmental performance.

Table 3. Peru: case studies on environmental management in the mining industry.

Firm	Reliance on external financing	Difficulties for technology transfer	Environmental behaviour of management	Location	Observations
			Large-scale mining		
SPCC	Centromin	Disruption in production process for introduction of new pyrometallurgical options (i.e., in-plant environmental solution)	• In 1992, Division of Environmental Affairs is set up • Independent environmental policy (does not follow any particular "environmental code of conduct" of its shareholders)	Copper smelter is 17 km from the port of Ilo	• 30×10^6 t/year of Cuajone and Toquepala tailings since 1977 (and half this in 1960–76) went directly into Ite bay, affecting water and the marine ecosystem • SO_2 in the off-gases of the smelter, under certain meteorological conditions, reaches the port and valley of Ilo • Disposal of smelter slag at the nearby seashore has affected beaches
Minero Peru	Centromin	Mineralogic complexity implies high costs and important in-house research for adaptation of new technology	• Developed a diagnosis of environmental problems at each production unit in the 1980s • Prepared a strategic plan in 1985 that included an environmental policy (firm ignored most of this later) • In 1991, prepared a PAMA for 1993–99 • In 1992, division of environmental affairs set up in La Oroya • Environmental master plan is being prepared with IDB support		• Conflicts with nearby communities re environmental care • Gases of the metallurgic centre affect the atmosphere (SO_2, CO, etc.) • Concentration tailings and mine waters seriously affect surface and groundwater (Mantaro river and others)

(continued)

Table 3 concluded.

Firm	Reliance on external financing	Difficulties for technology transfer	Environmental behaviour of management	Location	Observations
Shougang Hierro Peru [a]	Independent of multilateral international financing		• Cajamarquilla zinc refinery was constructed on the basis of environmental studies dating as far back as 1970 (almost a decade earlier) • Relies on a sulfuric acid plant to control SO_2 • Treats effluent in a plant • Office of mining hygiene and safety at each production unit is responsible for environmental-control tasks		• Contamination, under certain meteorological conditions and electricity-generation cuts, has affected a nearby observatory • Large amounts of tailings, after 1953, were directly disposed of at the seashore, apparently affecting the marine ecosystem
			Medium-scale mining		
Compañía Minera Milpo	Asking CAF for support for its 40 million USD expansion project		• Personal push from the general management for the environment had an important place in the general activities of the firm • Contracted a Canadian firm to design a 12 million USD tailings pond; constructed by local firms in the 1980s • Reforestation and development of back gardens in the mining camp • New expansion program includes a treatment plant for tailings-pond decanted water and a lime plant for neutralizing effluent		• Uses cut-and-fill mining • Back-filling of part of the tailings, with the rest going to tailings pond
San Ignacio de Morococha S.A.			N/A		N/A

Company				
Empresa Minera Iscaycruz		• In 1992, contracted an environmental consulting firm to develop an EIA of its production activities	• One of only two mines located on eastern side of Andes	• Uses back-filling and also has a tailings pond
Minera Yanacocha	• Negotiations with CAF, IIC, and BID for its 50 million USD Yanango electric-generating project • IFC is a shareholder of the project	• Environmental control has to be a responsibility of all • EIA developed by FIA (UNI) recommended integrating environmental activities of prevention and environmental control and making these the responsibility of the Department of Hygiene and Safety • Will abide by Peruvian regulations and also by Newmont's *Environmental Code of Conduct*, based on US environmental regulations • Contracted a US consulting firm to develop its EIA	• Cajamarca local council is asking for a canon, equivalent to 30–50% of total income tax	• Hydrometallurgic technology (Merril Crowe method) for gold production • Will use a three-layer carpet to prevent leakage of solutions

Note: BID, Banco Interamericano de Desarrollo (IDB); CAF, Andean Finance Corporation; EIA, environmental-impact analysis; FIA, Faculty of Environmental Engineering [UNI]; IDB, Inter-American Development Bank; IFC, International Finance Corporation [World Bank]; IIC, International Investment Corporation [IDB]; PAMA, Programa de Adecuación y Manejo Ambiental (environmental-management and adequation program); SPCC, Southern Peru Copper Company; UNI, Universidad Nacional de Ingeniería (national university of engineering) [Peru]; USD, United States dollars.
[a] Recently privatized.

SPCC has been under the same controlling interest, Asarco Incorporated (US), for 4 decades. The actual environmental effects of its production facilities have been bitterly disputed since its operations started in 1960. Only in December 1991 did the government and SPCC sign an agreement for a 300 million USD 5-year investment program (1993–97), of which one-third was to be used to partially control the operation's environmental impacts. This agreement granted SPCC a sort of environmental-regulation stability, as it was later excused from presenting a PAMA by the same legal instrument that introduced it and made it compulsory for all production units, that is, the Regulation of Title XV (Third Transitory Disposition). Only after the completion of this program will it be feasible for SPCC to proceed further in complying with the new regulatory framework.

Marcona iron mine has never been a matter of major public environmental concern. Nonetheless, it may be negatively affecting the marine ecosystem in its area, as it is a huge mine and has been operating its processing plants, particularly the concentration plant, for more than 3 decades. In fact, the mine has disposed of the enormous amount of tailings produced during all this time directly into the sea. The disposal of tailings into adequate tailing ponds would have prevented this environmental hazard. However, this provision was never made.

Marcona was developed and exploited in 1952–75 by Utha Construction in association with Cyprus Mines, both from the United States. The firm was nationalized in 1975 and remained a state-owned firm until December 1992, when it was bought by Shougang Corporation. Thus, the ownership of this mine, whether by well-known international firms or by the state, seems to have had no effect on the environmental issue.

However, under the new regulatory framework, its new owner, Shougang Corporation, is considering constructing an adequate tailings pond. This project will become even more urgent if the firm proceeds with expansion plans and doubles its production capacity to 20×10^6 t/year and adds a steel plant. Shougang Corporation is a property of the People's Republic of China, which may be considered a prospective NIE but certainly not a well-established developed country.

Centromín, SPCC, and Shougang Hierro Peru to a large extent use technology first introduced many decades ago. For example, two central processes with important implications for the environment in Peru are flotation concentration and the reverberatory furnace for copper smelting. Flotation concentration was introduced in the early 1920s; the reverberatory furnace, a century or so ago. Although important improvements in both technologies have provided major increases in productivity, these technical changes alone cannot alter the hazards that solid residues, effluent, and off-gases can cause.

For many decades, the industry has known about ways to control the tailings from the flotation process through specially designed ponds and treatment of the decanted water. This has for some time been on the university curriculum for mining engineers in Peru.

In the case of smelting with reverberatory-furnace technology, SO_2 has been difficult to control. This SO_2 needs to be at a minimum concentration of 4% to become economically recoverable by oxidation conversion into sulfuric acid. In the past 3 decades, and even earlier, new furnaces have been developed that integrate two or more pyrometallurgical processes to produce a high enough SO_2 concentration in the off-gases and are far more energy efficient. Inco Ltd, Noranda Inc., and Mitsubishi Minerals Corporation, among others, make use of this technology. Also, with CODELCO's modified converter, reverberatory-furnace technology offers SO_2 concentrations high enough to be economically converted into sulfuric acid, with greater energy efficiency and higher productivity.

Nonetheless, these are in-plant solutions, affecting the core installation of the process. As such, they are likely to involve major disruptions in production from the time setup until efficient operation is achieved. This is not the case with the control of flotation tailings, which is an end-of-line solution involving no in-plant disruption. In-plant disruption may have important costs over and above those of installing solutions, thus increasing the total net costs of environmental control.

The well-known international firms involved in large mining projects and the technical groups of the state-owned firms must have been aware of all these developments. However, Cerro de Pasço – Centromín, SPCC, and Marcona – Hierro Peru developed their mining and extractive metallurgical projects well before the environment became a serious issue in developed countries and for international multilateral agencies. Moreover, domestic environmental legislation presented important gaps and shortcomings that contributed to weak enforcement. This may explain why these firms made their first serious moves toward compliance with environmental regulations only as recently as the 1990s, motivated by more realistic and enforceable environmental legislation and by the strong support of the multilateral agencies.

On the other hand, state-owned Minero Peru developed its mining and metallurgical facilities in the 1970s and 1980s with foreign financing that in certain cases tied its support to adequate enforcement of environmental control. This company has shown relatively fair environmental behaviour throughout its period of operation. No one has filed serious complaints against Cerro Verde or the Ilo copper refinery. At Cajamarquilla zinc refinery, environmental assessments date back to 1970, long before Minero Peru began construction toward the end of that decade. Sudden disruptions of electricity flow, particularly as a result of sabotage, are the only isolated problems that have emerged, despite the mine's proximity to Lima (24 km).

We observed no major differences with respect to size, at least for the range of the firms we studied. Moreover, Milpo, a medium-scale firm, has shown concern for appropriate environmental management from the early 1980s, when it invested 12 million USD to construct a large tailings pond. This pond complements the use of some of the tailings to back-fill the mines. Furthermore, Milpo is looking for financing from multilateral agencies for a 40 million USD expansion and modernization program with a significant environmental component. This is planned for the next 2 years. The firm has enthusiastically supported a reforestation project in the areas close to the mining camp and also the development of ecological back gardens at the camp, where the miners' families grow their own vegetables. Ing. Augusto Baertl, the general manager, gave a decisive push toward this kind of environmental policy. He was convinced that environmental concerns had come to stay and that the industry must accept it. In the recent past, he was president of the National Association of Mining and Petroleum, which groups together private large-, medium-, and the more organized small-scale mining firms. This indicates that an influential manager who has a commitment to the environment may be a more important factor in a firm's environmental behaviour than ownership, size, timing of projects, or vintage of production technology.

The case of Milpo shows that, with determination, even a venture that started production in the 1940s can make important investments in environmental controls and maintain its competitiveness. This case also shows that longevity of the firm need not be a restriction on environmental performance.

Barriers to technological transfer: mineralogic complexity

The barriers to technological transfer stemming from mineralogic complexity may explain the attitude of some firms toward technical change. In-plant solutions to environmental problems may entail major changes to existing production facilities and processes. Such complexity seems to be the case with Centromín Perú and, particularly its La Oroya Metallurgical Centre. According to Centromín's manager of metallurgical operations, the complexity of the minerals processed at La Oroya can only be compared with that of minerals processed by Boliden in Suecia and Dowa Mining Company Ltd, and even so, La Oroya's is still the most complex (Huayhua 1993).

The bulk of the 250×10^9 t of minerals and concentrates treated at La Oroya comes from mines in the central Andes. Of this total, 150×10^9 t comes from Centromín's mines and the other 100×10^9 t comes from third parties. A central feature is that this plant processes "dirty" minerals and concentrates that many other smelters would be unwilling to process because of the complexity and high contents of toxic elements. La Oroya's third-party customers include some from Canada, the Philippines, and Spain, and the firm has also received offers from Russia.

Because of the materials Cerro de Pasço – Centromín has had to deal with for the past 7 decades, in-house innovation has been encouraged. In fact, the need for in-house research led the firm to set up the Department of Metallurgical Research at La Oroya in 1927. This department has since then been called "the school" of metallurgical engineers in Peru. J. Bonelli, director of the department for several years in the 1980s, and Ing. Agustín Mejía, head of Centromín's Division of Environmental Affairs (DEA), both stressed that mineralogical complexity was driving technological innovation at La Oroya and had implications for technology transfer (Bonelli and Mejía, personal communication, 1993[12]).

During La Oroya's long history, there has been an important process of incremental innovation. The opportunities (and difficulties) offered by mineralogic complexity have supported its longevity. Although this complexity implies substantial investments to adapt technologies originally designed to treat much simpler and more typical minerals, the net costs may be substantially reduced, or even turned into profits, by the much wider range of by-products.

Firms are likely to be more enthusiastic about incremental innovation or the introduction of a new production line than about radical innovation that disrupts production. Nevertheless, radical innovation is precisely what is needed in important sections of La Oroya's pyrometallurgical facilities, particularly to protect against atmospheric pollution. The situation is further complicated by the nature of its minerals and concentrates.

Thus, Cerro de Pasço – Centromín, in a long production life in which, until very recently, the environment was of little concern, has accumulated a huge environmental debit. According to International Management Centres (IMC) (London, United Kingdom), the valorization of Centromín, in preparation for privatization, included an estimate that compliance with relatively acceptable environmental standards would require at least 465 million USD in 1992. It should be noted that IMC's study was unavailable to us; therefore, we were unable to establish whether IMC included air-pollution control in that estimate. At Centromín, Ing. Mejía estimated the cost of air-pollution control at 500 million USD. In any case, the investment requirements are high; the costs of adapting technology and of disrupting

---

[12] Ing. J. Bonelli, former director, Department of Metallurgical Research, Centromín, interview, October 1993; and Ing. Agustín Mejía, head, Division of Environmental Affairs, Centromín, interview, 23 March 1993.

production are needed to give even a first approximation of the dimensions of the environmental task.

Compare these estimates with the 280 million USD cash and 60 million USD in eligible titles of Peruvian debt that were fixed as base price for bidding in March 1994 (Anon. 1994). Or compare them with Centromín's 400 million USD annual average sales for 1991–93 (Centromín 1992, n.d.).

It should also be pointed out that in 1992 the firm designed a PAMA on the basis of the IMC study and chose only those environmental-control projects with low investment and high pollution reduction per dollar. DEA produced a PAMA requiring only 45.2 million USD over a 7-year period, 1993–99 (Mejía, personal communication, 1993[13]). However, the PAMA omitted IMC's fundamental recommendations regarding atmospheric pollution. IMC explicitly recognized that the smelter emits 37 163 $N \cdot m^3$ SO_2/min, but the low concentration (0.68%) precludes economic recovery. But IMC added that this problem could be solved through technological innovation involving the production processes (Centromín 1992).

The approval of the PAMA granted the firm some regulatory stability: as far as environmental control is concerned, the new owners of Centromín must abide by the terms of the PAMA. Because this PAMA excluded atmospheric pollution, the likelihood of this problem being seriously addressed during the present decade is very slim.

In cases where minerals and concentrates are highly complex, such as at La Oroya, investment needed for transfer of environmentally friendly technologies is likely to be much higher than predicted. This cost is likely to become more critical when, as in the case of Centromín, a firm has been under major financial constraints and is unable to get fresh external resources because the country has a heavy foreign debt.

Reliance on international financing

Another important factor in a firm's environmental behaviour is reliance on international financing, particularly from multilateral development agencies. This has been observed in private, state, and foreign firms, independently of firm size:

- IDB was seriously considering an important loan to Centromín before the government decided to privatize the economy in 1991. The firm included a major environmental component at the express request of

---

[13] Ing. Agustín Mejía, head, Division of Environmental Affairs, Centromín, interview, 23 March 1993.

IDB, whose technicians had visited La Oroya and had apparently been shocked by its situation. Later on, IDB helped Centromín prepare for privatization, commissioning the IMC study to evaluate the firm's environmental debit. Furthermore, and particularly relevant here, IDB was helping the firm prepare the terms of reference for a 2 million USD study to produce an environmental master plan. The study would cover the global problem of Centromín's impact (on La Oroya as an urban centre, on its are of operations, and on local agriculture) within a medium- and long-run perspective.

• Buenaventura has kept important long-term contacts with the IFC of the World Bank and the Inter-American Investment Corporation (IIC) of IDB. Thus, IIC, the Andean Finance Corporation (CAF), and the organization of the Andean Pact are helping to finance Buenaventura's Ishihuinca gold mine.

• SIMSA has approached the IIC and CAF for support for the50 million USD electricity-generating Yanango project. The particular importance of this domestic firm, which is the largest private zinc producer in the country, may be gathered from the fact that it may participate as a bidder in the privatization of Minero Peru's 101.5×10^9 t/year Cajamarquilla zinc refinery (Centromín n.d.).

• Milpo, the third largest medium-scale firm, has been looking for support at CAF and other multilateral agencies for its 40 million USD expansion plan.

• SPCC received a 60 million USD grant from CAF for the development of its Toquepala and Cuajone marginal-minerals leaching project, which is included in its 300 million USD 5-year investment program.

International financing in the new greenfield projects is even more evident:

• Iscaycruz had by May 1993 progressed well in the negotiations for financing with CAF and IIC (Bressi, personal communication, 1993[14]). It may be recalled that this firm is owned by Paraibuna Metais (Brazil)

---

[14] Ing. Rodolfo Bressi, general manager, EME Iscaycruz, interview, 28 May 1993.

(45%), in association with Buenaventura, Minero Peru, and Marc Rich (Switzerland).

- IFC has participated in the financing of Newmont's majority-owned Yanacocha gold project and is at present a shareholder.

Lastly, worth noting although it is not one of the case studies, is Mantos Blancos's Quellaveco copper deposit, which is of comparable type and size to those of Toquepala or Cuajone. The firm was undertaking a 2-year feasibility study for the deposit's eventual development, and IFC was participating as a shareholder in the project.

The pressure that the multilateral agencies, particularly the World Bank and IDB, have exerted on firms' environmental behaviour in recent years has been important. This has been explicitly recognized by most of the firms mentioned. However, an important exception to this observation would be Shougang Hierro Peru, which, as it has been pointed out, is a property of the People's Republic of China and, as might be expected and is in fact confirmed by the firm (Alfaro, personal communication, 1993[15]), works outside the spectrum covered by these multilateral agencies.

Response to the new environmental regulatory framework

The analysis of the case studies indicates a particular way large mining have tended to relate to the new environmental regulatory framework. Vizcarra (1982), noted that many firms exert direct influence through governmental or parliamentary commissions set up to investigate issues of public concern. In fact, the importance of these firms in the national economy may account for this approach.

In general, enforcement of the environmental regulatory framework has been very weak. The new legislation, emerging in the period following the publication of the new Code, significantly departs from the previous regulatory system. The emphasis in the new regulatory framework is on providing firms with ways to progressively comply with appropriate environmental standards, rather than merely penalizing them for failure to meet these standards. For this reason, we postulated the likelihood of a much higher rate of compliance.

The case studies confirmed this postulate. At a formal level, this might be indicated by the emergence during the 1990s of specific environmental offices in the organization charts of several of the analyzed firms. Moreover, in all cases, third parties were performing environmental research, according to law, to produce

---

[15] Ing. J.Cl. Alfaro, technical manager, Shougang Hierro Peru, interview, 28 December 1993.

the firms' PAMAs or EIAs. This work was, in general, coordinated by the production departments and the specific environmental office or with the office of safety and hygiene. The environmental office, or the one in charge of these activities, in most cases is just one step down from operations management and only two from general management. People at the high decision-making levels can thus be rapidly informed about environmental developments, reflecting a clearer focus on environmental issues.

As it might have been expected, the large firm in most cases allocated important technical resources (professional groups and laboratory infrastructure) of the firm to support the work of external environmental consultants. Thus, as it has been noted, Centromín produced its PAMA on the basis of the IMC valorization study. It should also be mentioned that IMC had subcontracted a US environmental auditing unit to develop the environmental component. All production departments and the Division of Metallurgical Research supported those producing the PAMA.

Minero Peru has not confronted relevant environmental problems in its units. As of 1990, it had developed environmental activities only at its Cajamarquilla zinc refinery (Vidalón, personal communication, 1990[16]). However, we learned that by June 1993, environmental-monitoring activities had begun at its other two production units — Cerro Verde copper mine and Ilo copper refinery.

The Environmental Control Office at Cajamarquilla was set up when the plant began producing in the early 1980s. It has maintained a permanent monitoring system for its emissions, liquid effluent, and solid waste. Thus, it analyzes, for example, air-pollution control, at 17 permanent stations; the office also works with Sedapal S.A., Lima's water firm, to monitor the Rimac river water, as well as other control tasks. The analyses are done with the assistance of two chemists, a biologist, seven supporting samplers, and the staff of the plant laboratory.

In 1992, Minero Peru, like Centromín, had IMC do its valorization in preparation for privatization. This time, IMC subcontracted Morgan & Grenfell (United States) to develop the environmental component of the study. Although IMC's reports were unavailable to us, interviews at the production units and central offices indicated that only specific aspects, mainly of the environment in the workplace, were pointed out for correction. in this case, the mining firm's participation significantly contributed to the work of the environmental consultants.

Since the mid-1970s, SPCC has commissioned environmental studies. The first one was to evaluate the potential environmental impacts of developing its Cuajone mine to more than double its capacity. Another study several years later

---

[16] Ing. J. Vidalón, head, Metallurgic Direction, Minero Peru, 26 November 1990.

was undertaken to respond to a parliamentary commission set up to investigate the environmental impacts of the firm's operations in its zone of influence. More recently, in the 1990s, the firm commissioned studies to prepare its 300 million USD 5-year investment program, which includes 100 million USD for environmental projects. For its mid-1970s and mid-1980s studies, the firm contracted Dames & Moore (United States). In the 1990s, it contracted Rescan Environmental Services Ltd (Vancouver, BC, Canada) to study the feasibility of disposing of Cuajone and Toquepala tailings in the sea; and Klohn Leonoff Ltd (Richmond, BC, Canada) to study an alternative method for disposal on the mainland.

SPCC set up its Environmental Protection and Research Centre in Ilo to undertake environmental activities at its operations. However, we were unable to find out specifically what this centre is working on. It may be added that the firm also set up in Lima a Directorate of Environmental Services, which is in charge of the firm's relations with official environmental-control offices and commissions, as well as coordinating the work of the consultants. SPCC is relying fundamentally on well-known foreign firms, especially from the United States and Canada, to develop its environmental activities.

On the other hand, Shougang Hierro Peru contracted LAGESA to develop its PAMA. LAGESA, a domestic consulting firm, was one of the case studies we used to analyze Peru's technological capabilities. Similarly, Iscaycruz, the polymetallic greenfield project controlled by Compañía Paraibuna de Metais, commissioned FIA at UNI to undertake its EIA.

The two firms controlled by foreign investors from NICs (Shougang Corporation [Beijing] and Compañía Paraibuna de Metais [Brazil]) have, independent of their size, contracted domestic environmental technological capabilities.

Of the three medium-scale firms, Buenaventura is the only one that set up a consulting firm — BISA. This firm has mainly worked in mining and metallurgy and has been registered at MEM to perform EIA in these areas. However, for environmental studies it has expressed a willingness to look for domestic or foreign partners to complement its technological capabilities.

Milpo and SIMSA, the other two medium-scale firms, have relied, to an important extent, on foreign consultants for its environmental activities. Milpo contracted Golder Associates of Canada to design its 12 million USD tailings pond, although the construction was commissioned to domestic contractors. However, its much smaller projects of reforestation and back-garden development were assigned to a local nongovernmental organization, Friends of Peoples Close to Nature.

Similarly, SIMSA contracted Tecno Serv (United States) to develop its EIA. However, this firm has relied mainly on the local infrastructure of laboratory services (such as Universidad Nacional Agraria, Universidad Nacional Mayor de San Marcos, and the National Institute for Agricultural and Industrial Research).

In rough terms, it may be concluded that the medium-scale firms do not seem too keen to take on much of the environmental-control activities themselves but prefer to employ consultants and external services for this purpose. Also, the two new greenfield projects, Yanacocha and Iscaycruz, have shown that from the very beginning, they have integrated the environmental dimension into their design on the basis of EIAs commissioned to consulting firms. As already mentioned, Newmont, the controlling interest of Yanacocha, was the only firm that expressed its willingness to abide both by the Peruvian regulations and the much stricter regulations of the United States.

CHAPTER 5

FORMAL AND *GARIMPO* GOLD MINING AND THE ENVIRONMENT IN BRAZIL

Maria Hanai

For the past 20 years, the environmental impacts of mining have been a growing concern. In Brazil, this concern was reflected in the passage of the 1981 National Environmental Policy. This legislation consolidated existing regulations and created an administrative structure to implement them. At about the same time, a major expansion occurred in both formal and informal gold mining. The emergence of the *garimpo* (artisanal) phenomenon, particularly in the informal sector, was primarily due to the dramatic increase in the gold price in the late 1970s and to the high unemployment at that time.

This paper provides an overview of the gold-mining sector in Brazil. It examines the environmental effects of the technologies used, identifies factors in the environmental behaviour of formal gold-mining companies, and suggests how the regulatory regime can be enhanced to minimize the effects of gold mining on the environment.

This research was done from April 1990 to October 1992. It involved the use of both secondary sources (reports, articles, books, and statistics) and direct interviews with government officials, nongovernmental agencies, mining-company executives, and prominent members of the *garimpo* community.

History

Although gold was known to be present in Brazil in the 16th century, commercial exploitation only began a century later with the discovery of rich deposits in the states of Minas Gerais, Goiás, and Mato Grosso. In the 18th century, production in these areas (and others) made Brazil the largest producer of gold in the world. During this time, output was 830 t, or 58.4% of the global output, which was about 1 421 t (Berbert 1988).

Table 1. Investments in exploitation of gold and mining, 1981–90. [a]

Year	Gold (million USD)	(%)	Mining [b] (million USD)
1981	50.2	25.1	200.0
1982	52.8	24.1	218.8
1983	31.2	23.3	133.8
1984	68.5	39.4	173.0
1985	50.0	41.4	120.9
1986	34.4	39.2	87.8
1987	60.5	31.4	192.4
1988	87.8	65.8	133.4
1989	48.2	55.4	87.0
1990	30.0	50.0	60.0

Source: DNPM–DEM (n.d.), SIPEM Project, Mineral Yearbook.
Note: USD, United States dollars.
[a] Constant price, 1990+ 100, obtained on basis of the current values in national currency, converted at the official rate of exchange (indexed by the US General Price Index).
[b] Excluding petroleum and natural gas but including gold.

However, production declined from the middle of the 18th century, mainly because the rich alluvial deposits of Minas Gerais were exhausted. Technical problems related to the development of underground mines contributed to this decline. After Brazil's independence in 1822, the country was opened to foreign investment, and about 50 gold-mining companies were established. Most were unsuccessful; by 1888, the beginning of the republican period, only four remained in operation. Of these, only one, St John d'El Rey Mining Company, is still in operation; it is now named Mineração Morro Velho S.A. (Maron and Silva 1984).

In 1931–1980, Brazil produced only 305 t of gold, about 0.6% of global production (Berbert 1988). However, this situation changed with increasing gold prices and the collapse of the Bretton Woods agreement. With the ensuing gold rush — the greatest in Brazilian history — Brazil rejoined the world's major gold producers.

The development of the formal gold-mining sector

As a result of the gold rush, the investment in mining exploration and development in the formal sector in 1981–90 (Table 1) totaled 1.407 trillion United States dollars (USD), or about 20% of the total mineral-sector investment in that period (excluding petroleum and natural-gas investment) (DNPM–DEM n.d.).

Table 2. Estimated industrial (formal) gold production by economic group, 1993.

Economic group	Production (t)	(%)
Compañía Vale do Rio Doce	134	30.0
Bozzano Simonsen – Inco–TVX[a]	131	29.5
RTZ–Autram–Cobem	55	12.4
Amira Gencor	45	10.1
Others	80	18.0

Source: DEM–DNPM (n.d.).
[a] In June 1993, Inco sold its controlling share of TVX Gold, whose chair is based in Brazil (Anon. 1993; Haliechuck 1993).

These investments in formal mining resulted in the discovery of about 25 gold deposits with economic potential (Mackenzie and Dogget 1991). Several new companies were formed to develop these, principally Compañía Vale do Rio Doce (CVRD) (1984), São Bento Mineração S.A. (1986), Rio Paracatu Mineração S.A. (1987), and Marex Mining (1987). The old Mineração Morro Velho S.A. also increased its capacity and formed two new companies, Jacobina Mining and Commercialisation (1983) and Serra Grande Mineração (1984). After the expansion of the 1980s, a period of consolidation occurred. By 1991, the sector had 27 companies; a number of these belonged to one of four economic groups:

- The Brazilian state-owned company;

- Bozzano Simonsen Group, associated with the South African Anglo American Corporation and the Canadian Inco–TVX incorporation;[1]

- RTZ (United Kingdom) – Autram-Cobem; or

- Amira–Gencor (South Africa).

These groups (Table 2) were responsible for 82% (365 t) of the total formal-sector gold-mining output of Brazil in 1993.

---

[1] In June 1993, Inco Ltd sold its control block in TVX Gold, which holds interests in six gold and silver mines in Canada, the United States, and South America, including Brazil. TVX Gold was created in 1991 through the merger of gold assets held by Inco, and after the sale, TVX Gold's chair, Eike Batista, who is based in Brazil, became the largest single shareholder, with 13% stake (Haliechuck 1993; *Wall Street Journal* 1993).

Labour force in formal mining

In 1989, the formal mining sector in Brazil directly employed 8 744 workers (excluding subcontractors). In contrast, there were about 350 000 *garimpeiros* (prospectors) at this time. Technologies that economize on labour help to explain the decreased number of workers in industrial mining. For instance, in 1985 the output of Mina Grande, one of the oldest mines of Morro Velho, was 0.75 t of ore per worker's shift. By contrast, in 1982, the new mines of Cuiabá–Raposos, which are part of the Morro Velho mining group, produced 4 t of ore per worker's shift in surface operations and 3 t of ore per worker's shift in underground operations (Brasil Mineral 1985).

In businesses that count on the participation of outside capital, the new technologies in the production process are usually introduced by the foreign partner. The Cuiabá–Raposos mines, already cited, is an example. To drill deep shafts, the company replaced its manual drilling machines by modern equipment that reduced the number of workers employed and the time required to excavate the gold. In this case, shaft-sinking technology was introduced by Shaft Sinkers Ltd of South Africa, a subsidiary of the joint-venture partner, Anglo American (Minérios: extração e processamento 1991).

Health and safety

Underground conditions in the older mines leave much to be desired, with high injury rates, poor ventilation, and evidence of high rates of silicosis among the workers. According to the Miners Union of Nova Lima (state of Minas Gerais), where the largest mining complex of Morro Velho is located, about 40 accidents occur every month; in the 1980s, about 80 miners died from work-related accidents. According to the Institute of Social Welfare in Nova Lima (Morro Velho's headquarters), 4 miners die from silicosis and 13 new cases are reported every month, on average (Ribeiro 1990).

The problems of occupational health and safety have been a source of tension between workers and management, principally at Morro Velho, whose old mines provide precarious health conditions. Mining companies are more interested in increasing their profits than in taking responsibility for directly managing the mass of workers. This attitude has led the mines to use subcontractors for the manual labour, principally in the mining phase, which has some operations similar to those in civil construction. For example, in Rio Paracatu, 62% of the firm's 1 152 workers were subcontractors.[2]

---

[2] (interview at Rio Paracatu Mineração, 1991.

Environmental

The Rio Paracatu mining operation is a case of environmental awareness coupled with financial success. At this open-pit mine, 0.6 g Au / t is recovered by using the carbon-in-leach technique.

However, the more refractory ores of São Bento led the principals (Amira and Unamgem Mining and Metallurgy) to adopt sophisticated technology developed by their joint-venture partner, Gencor (of South Africa). The firm has adopted both pressure oxidation and biological oxidation techniques, and once their teething troubles are overcome, these processes should give only minor environmental problems (Reis et al. 1991).

Some firms have also adopted new technology for the cyanidation step — the carbon-in-pulp and heap-leaching systems. Among the firms using these techniques are Rio Paracatu, Morro Velho, CVRD, Xapetuba Mining, Jacobina Mining and Commercialisation, and Serra Grande. In every case, closed-circuit flowsheets are used to ensure environmental protection. In addition, electronic monitoring and control equipment has been installed in a number of these plants, further strengthening environmental security.

Garimpo gold mining in Brazil

The leap in the gold prices at the end of the 1970s led to an upsurge in informal mining activity exceeding that of the gold rush of the previous century. Up to this time, informal mining had been limited largely to the Tapajós district (in the Amazon region to the south of the state of Pará). In 1973, only 5.9 t of gold was produced by the *garimpeiros*, and 5 years later this had increased to 18 t (Araújo Neto 1991).

The rate of *garimpeiros'* discoveries increased greatly after this time, with alluvial gold deposits discovered in Alta Floresta, in Mato Grosso (1979); Rio Madeira, in Rondônia (1979); Serra Pelada, in Pará (1980); Cumaru and Tucumã, in Pará (1981); Apiacas and Juruena, in Mato Grosso (1981); Catrimani and Uraricoera, in Roraima (1987); and several other sites (Mackenzie and Dogget 1991). Additionally, output considerably increased at existing *garimpos* sites, such as Paruari and Tapajós, in Pará; Gurupi, in Maranhão; and Lourenço, in Amapá.

Parallel to this upsurge, the amount of gold sold on the black market increased disproportionately. Table 3 shows how the official and unofficial outputs diverged after the mid-1970s; the two outputs corresponded only from 1990 on, following the flotation of Brazilian currency (DEM–DNPM n.d.).

Table 3. Gold output of *garimpos*, 1973–90.

Year	%	(A) Official (t)	(B) Estimate (t)	Difference (B÷A) (%)
1973	1.2	5.9	4.7	79.7
1974	1.1	9.0	7.9	87.8
1975	1.5	9.6	8.1	84.3
1976	2.5	9.9	7.7	77.8
1977	1.6	12.1	10.5	86.8
1978	5.4	18.0	12.6	70.0
1979	1.1	31.7	30.6	96.5
1980	9.7	35.9	26.2	73.0
1981	12.9	37.6	24.7	65.7
1982	20.9	41.0	20.1	49.0
1983	47.5	63.6	16.1	25.3
1984	30.6	55.0	24.2	44.4
1985	22.2	65.0	42.8	65.8
1986	14.8	75.0	60.2	80.3
1987	22.7	78.0	55.3	71.0
1988	34.3	90.0	55.7	61.9
1989	29.5	80.0	50.5	63.1
1990	—	55.0	55.0	—
1991	—	41.9	41.9	—

Source: 1973–80, Araújo Neto (1991); 1981–91, DEM–DNPM (n.d.).

A number of researchers attributed this boom principally to socioeconomic factors (Salomão 1984; Cota et al. 1986; Susczynski 1988). They considered rural-settlement policies and the failure of agrarian reform as being key factors. The situation was exacerbated by the series of economic crises that struck Brazil after 1973 (the first oil shock) and worsened in 1979–83 (the second oil shock). A further factor was extensive highway construction in the 1960s and particularly in the 1970s, which facilitated migratory movements to the Amazon region, where *garimpo* activity is concentrated. Such highways included the Belém–Brasília, the Transamazônica, the Cuiabá – Porto Velho, the Cuiabá–Santarém, and sections of the Perimetral Norte.

The introduction of new technology and its social effects on *garimpos* operations

In addition to the opening up of new areas for production, the surge in *garimpo* output in the late 1970s was driven by new technologies. Thus, *garimpos* operations became more intensive, as well as more extensive. In particular, pumps were introduced, as were various types of crushers.

This mechanization, albeit limited, meant that a new category of miner appeared: an entrepreneur who had the capital to purchase and maintain the machinery. Such people became site bosses, and the wealth they accumulated gave them much economic, political, and social clout. By 1988, they were powerful enough to influence the wording of the new Constitution; indeed, by 1990, the *garimpeiros* represented a significant voting block, particularly in the Amazonia region.

As a result, for the first time, artisanal mining was addressed in the Constitution, and *garimpo* activity ceased to be seen as an illegal form of mining exploitation. It was defined according to the mineral sought and the type of deposit. The Constitution also assured the miners the mineral rights to the deposits they were exploiting (art. 174, para, 4, regulated by law No. 7805/89). It also established the right to form cooperatives with a status equivalent to a formal mine (art. 174, para. 3).

The environmental effects of *garimpos* activities

The technical improvements secured by the *garimpeiros* were all directed at mining more ore; the gold-recovery processes were still limited to gravity concentration and amalgamation. However, the environmental effects of this modest mechanization were severe, with deforestation, siltation, and land sterilization occurring over wide areas (Silva et al. 1989; Silva 1991). As by-products of this increased activity, the incidence of infectious diseases, such as malaria, increased greatly among the indigenous people living near the *garimpos*, in particular the Yanomamis. This tribe has inhabited the region north of the state of Roraima for several thousand years, and contact with the *garimpeiros* has led to a serious corruption and decline of their culture.

According to medical examinations done by the National Health Foundation of Roraima between January and May 1991, 50.3% of the 9 588 Yanomamis examined had malaria; 25% of these had virulent forms for which no satisfactory cure has been found. The *garimpeiros* suffer as much as the indigenous people from this disease. According to data from the Brazilian Ministry of Health (1990), more than 99% of the cases in Brazil occur in the Amazon region, with 70% of these in the 30 districts where *garimpo* activity is ongoing. Of the 560 143 positive blood smears, 376 554 were from the inhabitants of these districts (DMMH 1991).

Increased gold output means increased use of mercury. No direct measurements of use have been made, but very little mercury is recovered from the amalgamation process the *garimpeiros* use (despite the availability of simple and cheap technology). Hacon (1990) compiled the results of Brazilian studies (Lacerda et al. 1987; Martinelli et al. 1988; Couto et al. 1989; Pfeiffer et al. 1989; Malm et al. 1990) and found levels well above the limits established by the World Health Organization and Brazilian public-health agencies. The studies also revealed that mercury contamination reached not only the population groups endogenous to *garimpos* sites but also exogenous groups, an indication of the extent of the contamination.

Environmental regulations

Although Brazil has had environmental regulations since the 1930s, they were not consolidated into a general, formal code. However, during the late 1970s, when gold production increased rapidly (Table 4), environmental legislation increased correspondingly. This stage began in 1981, when the National Environmental Policy was approved. This law was presented in detail in June 1983. The administration of this law and the promulgation of environmental laws by the states was dealt with by the Sistema Nacional de Meio Ambiente (SISNAMA, national system for the environment) (law No. 6938/81 and decree No. 88351/83).

At both the state level and the federal level, new environmental laws that dealt with the relationship between mining and the environment were passed. At the federal level, the Comisión Nacional del Medio Ambiente (CONAMA, national commission for the environment) — the consultative and deliberative body of SISNAMA — approved a number of important resolutions. These required presentation of an environmental licence issued by an acceptable environmental authority. This in turn required companies to undertake estudios de impacto ambiental (EIAs, environmental-impact studies) and relatórios de impacto ambiental (RIMAs, environmental-impact reports) (CONAMA resolutions No. 01, 23 January 1988, and No. 09, 16 December 1990).

The 1988 Constitution also set out guidelines governing the relationship between mining and the environment: these provisions included the obligatory preparation of EIAs before initiation of mining or activities with the potential to cause significant environmental degradation (art. 255, para. 3, item IV). They also required *garimpo* mining operations controlled by cooperatives to take due account of the environment (art. 174, para. 3).

Table 4. Brazilian formal and *garimpo* production, 1973–91.

Year	Formal production		*Garimpo* production	
	t	%	t	%
1973	5.1	46.4	5.9	53.6
1974	4.8	34.8	9.0	65.2
1975	3.9	28.9	9.6	71.1
1976	3.7	27.2	9.9	72.8
1977	3.8	23.9	12.1	76.1
1978	4.0	18.2	18.0	81.8
1979	3.3	9.4	31.7	90.6
1980	4.1	10.0	35.9	90.0
1981	4.4	9.8	37.6	90.0
1982	4.6	10.1	41.0	89.9
1983	6.2	8.9	63.6	91.1
1984	6.7	10.9	55.0	89.1
1985	7.6	10.5	65.0	89.5
1986	9.3	11.0	75.0	89.0
1987	13.1	14.4	78.8	85.6
1988	22.2	19.8	90.0	80.2
1989	22.9	22.3	80.0	77.7
1990	29.9	45.2	55.0	64.8
1991	34.0	44.8	41.9	55.2

Source: 1973–77, Araújo Neto (1991); 1978–91, DEM–DNPM (n.d.).

In 1989, regulations on art. 174, paras. 3 and 4 of the Constitution were issued, creating the Garimpo Mining Permit (law No. 7805, 18 July 1989). This, among other provisions, made the granting of mining rights dependent on submission of the environmental licence to the appropriate environmental authority (arts. 2 and 18). It also made *garimpos* mining activities that damaged the environment subject to a series of penalties (arts. 20 and 22).

At the state level, some units began to define specific regulations and to create state environmental agencies, following directives of the National Environmental Policy. Among the gold-mining states, Minas Gerais — the state with the traditional mining region — was the first to set up structures for managing the environmental problems related to mining. There, beginning in 1981, a series of decisions of the Conselho de Política Ambiental (state council for environmental policy) led to regulations governing mining activities with the potential for damaging the environment. Of all the gold-producing states, Minas Gerais has the most controls on environmental impacts of mining. (It also produces the highest volume of industrial gold: 15.4 t, or 51% of Brazil's total production of industrial gold in 1990.) The formal mines and the *garimpeiros* differ in their response to these regulations, for reasons set out below.

An analysis of the environmental behaviour of formal and *garimpo* gold mining

There is clear differentiation between the environmental impacts caused by formal mining and those caused by *garimpos* activities. The formal mining, despite having the potential to cause serious occupational health and safety problems (mainly when the miners are underground), damages the environment less than *garimpo* mining. The more predatory behaviour of the *garimpeiros* can be attributed to various factors, discussed in the following subsections.

Scale of operations

The main businesses in Brazil's formal mining sector are large state firms and firms with private national and foreign investors. Large mining companies exploiting a highly profitable mineral will have the resources to adopt good environmental practices.

Such companies more often have a better environmental profile than the *garimpeiros* and are familiar with the technologies involved: for example, Rio Paracatu, connected with the British conglomerate, Rio Tinto Zinc, and the Brazilian state-owned CVRD (Birkin 1989; *Mining Journal* 1990).[3] São Bento, connected with the South African firm Gencor, has adopted Gencor's environmental practices.

Besides investing in methods for minimizing environmental degradation, some businesses have adopted compensatory environmental strategies, such as the creation of reserves of fauna and flora in areas they own. CVRD has also invested in improvements to community installations to it gain better public acceptance.

In contrast to the big formal operations, the *garimpos* sites are of a small or medium scale. They are financed by native investors, who are largely unconcerned about the environment and are without the same motivation for adopting practices less damaging to it. In addition, the majority of the site bosses operate

---

[3]   Also, interviews at Departmento Nacional de Produção Mineral (DNPM, national department of mineral production) and CVRD.

in regions where they do not own the property on the surface and therefore have little interest in investing in the conservation and protection of this asset.

Company locations versus *garimpo* mobility

Formal gold-mining operations are planned to last many years and require lots of capital. This makes these mines visible and accessible to control and supervision by the social elites in the country, by the communities, and by the government and nongovernmental environmental agencies. A good number of these mines are near urban areas that in many cases predate the projects, so they are more vulnerable to community pressure and social control. They are also subject to more control because they are in states where government environmental agencies are more structured, as in Minas Gerais. In addition, formal gold-mining projects are subject to more social control because they are large and in the majority of cases connected with large economic conglomerates. Various denunciations by the community have been registered with the environmental agencies.[4]

Supervision and control of the *garimpos* sites are much more complex. The locations are numerous (estimated at 2 000 in just Pará) and difficult to reach, and this activity is distinguished by its mobility. The mobility is reinforced by the small size of the deposits (the work at each site is generally of short duration), by the small investments required to mount the infrastructure directly at the mining site, and by the ease with which machines and equipment can be transported from one site to another.

Also, the majority of the sites for *garimpo* gold mining are in the Amazon region (which produced 79.5 t of *garimpo* gold in 1989). In this region, government environmental agencies are not well equipped to control the pollution, and at times the voices of the victims of pollution are confused with those of the polluters themselves. This is because the *garimpeiros* and the people living in the gold-prospecting areas are both harshly affected by the environmental pollution the mining provokes. The effects of pollution are also strongly felt in the communities that tend to spring up around the *garimpos* to provide the social and the economic activities needed to develop this activity.

Technological dualism

Most of the formal mining companies began to operate in the 1980s and were able to use technologies with high economic performance, developed mainly in the 1970s. These technologies basically increased productivity by reducing the use of

---

[4] Interview at Fundação Estadual do Meio Ambiente (FEAM, state foundation for the environment), 1992.

manual labour and inputs in general, and their introduction was accompanied by a parallel development of techniques and processes to minimize environmental impacts. These impacts were related to the use of dangerous products and to the release of toxic elements (contained in the mineral) during the treatment process.

The formal mining companies' adoption of methods less damaging to the environment was also bolstered by the possibility of incorporating some technologies to reconcile the best economic performance with the best environmental performance. For instance, SO_2 from roasting plants is collected to produce sulfuric acid. Although the market for this acid is not always favourable, its production helps to reduce the cost of environmental protection (Morro Velho). Another example is the technique of backfilling, which makes possible the best use of solid wastes. After the ore is processed above ground, the waste material is returned to the mined-out areas. This reduces the expense of underground timbering and provides an alternative to leaving the waste material in tailings dams (São Bento and Morro Velho).

The use of the pressure-oxidation autoclave, which uses compressed oxygen (harmless to the environment) as its basic input, and of biotechnology for the metallurgical process is currently small scale, as is the use of most of the technologies that, with regard to the specific characteristics of ore exploitation, reconcile good economic performance with good environmental performance (São Bento) (Brazil Mineral 1985; Ribeiro 1989).[5]

According to specialists who work in formal gold mining, the incorporation of new technologies is not brought about through a technological dependency; rather, assimilation of the technology by domestic technicians is related to the existence in Brazil of good educational institutions for mining.[6]

Rio Paracatu, besides incorporating more ecologically sustainable technological processes, adopted new administrative practices. Environmental management (including management of occupational health and safety) is now shared among all employees, as well as the manual labourers (who are subcontractors).[7]

The *garimpos* mining activities, on the other hand, developed in a labour-intensive and highly polluting framework, with technologies that demanded little capital. Because the *garimpos* are small- and medium-scale undertakings, they expand by using mechanical and semimechanical technologies for rapid and easy extraction of gold, with little or no attention to the efficiency of the operation or to

---

[5]   Also, interviews with the director of São Bento and technicians at Fundação Centro Tecnológia (CETEC, central technological foundation) and FEAM, 1991 and 1992.

[6]   Interviews at DNPM and FEAM and with independent consultants in the mining sector.

[7]   Interviews with technicians of Rio Paracatu, 1991.

environmental management. The very organizational structure of *garimpos* activities is also a serious obstacle to the introduction of more suitable technologies. For instance, difficulties for environmental control are created by the existence of a variety of operational systems in the same gold deposit, with a variety of production technologies (dredges, mills, high-powered hoses, etc.) managed by different *garimpos* bosses.

Environmental regulations

Until 1983, only four industrial gold-mining companies operated in Brazil, and only one, Morro Velho, was dedicated exclusively to the production of this metal. The other firms produced gold as a by-product of the mining and refining of other minerals (Maron and Silva 1984).

Today, about 30 companies are operating, and almost all began when the a new set of environmental regulations, directed at various sectors of economic activity, including the mining sector, was introduced. Consequently, only Morro Velho, which had operated with conventional methods and without environmental concerns, had to restructure itself to comply with these environmental requirements. The new firms on the national scene could incorporate technologies to meet the environmental regulations at the beginning of their operations. These firms had reduced expenses for environmental control because the cost of prevention is less than the cost of repairing damage.

The businesses were strongly motivated to follow the environmental regulations because they were essentially obligated to: the legal requirement to present the environmental licence (issued by the appropriate environmental agency in DNPM) to receive the rights to mine and to begin operations pressured the firms to carry out EIAs and RIMAs. Furthermore, some financing agencies, such as the National Economic and Social Development Bank, began to require environmental licences before giving credit to the companies, which broadened the range of environmental supervision.

Before the implementation of the new environmental regulations, basically only one mining business operated in Brazil, Morro Velho, which had been in existence for more than a century. In the process of refining the gold ore, this company released toxic gases (SO_2), destroying all of the vegetation downwind of its chimneys. The firm also threatened the water supplies because it had no type of holding dams for the rejected materials (Maron and Silva 1984). Only in 1980, with increasing environmental and community pressure, did Morro Velho begin to adopt methods of environmental control. These included dams for holding the rejected material, a new refining plant, and a plant to manufacture sulfuric acid (to avoid the release of SO_2 into the atmosphere). In addition, the company carried out programs to monitor the quality of air and water and a deposit of slimes containing arsenic (Brazil Mineral 1989).[8]

Despite these measures, Morro Velho still carries heavy environmental liabilities. Because it used conventional technologies and paid no attention to the environment for more than 100 years, serious problems are still unsolved. For example, for decades the company deposited waste materials with high levels of arsenic alongside the Cardoso river. With time, the material compacted, and now it is around and under urban buildings.[9]

The introduction of the new environmental regulations had little impact on the investment decisions of the other industrial mining companies, though, as most of these were just entering the market when these regulations were being adopted. The increase in gold prices that occurred at the same time was responsible for the expansion of these companies.

Although these environmental regulations had little effect on company investments, they did affect organizational structure and the configuration of the companies' mining and processing projects. All of these companies contracted their own personnel or outside consultants specialized in environmental issues, and the question of the environment came to be seen as a key element to be incorporated into gold-mining projects.

The new environmental regulations had very little impact on the *garimpos*, in contrast, despite the fact that these prospecting sites were also obligated — even by the Constitution — to adopt measures to preserve the environment. Such measures were never adopted by the *garimpos*, however, because of the very nature of their activities, such as

- The short period between initial operation and exhaustion of deposits;

- The mobility of *garimpeiros*;

- The disregard for obtaining environmental licences as a requirement for credit (the *garimpeiros* operate within an informal economy);

- The expansion of *garimpos* into areas independently of the granting of formal mining and implementation rights;

---

[8]   Also, interviews with technicians at DNPM, CETEC, and FEAM, 1992.

[9]   Interview with technicians at FEAM, 1992.

- The heterogeneous nature of the mining techniques used in a contiguous area and the consequent difficulties in standardizing technologies and environmental recovery measures;

- The lack of dissemination of nonpolluting technologies;

- The lack of technical assistance; and

- The unsuitability of environmental legislation, which imposes the same environmental controls on *garimpos* as on large-scale mining companies, despite the evident differences between the two.

Final comments and suggestions

The environmental degradation generated by gold mining is concentrated in the *garimpos* areas, where problems of pollution have reached significant proportions and there is little indication of change in the near future.

The *garimpo* workers are more predatory with regard to natural resources than the formal-sector mining companies are. The principal reason for this is that the large corporations have much more capital and can more easily absorb the cost of advanced technologies for gold production. The *garimpeiros* have much less capital to adopt environmental technology. This problem is intensified by the fact that these workers are seldom qualified (but are just people who have escaped the widespread unemployment in the country), in contrast to the highly qualified people employed in formal gold mining.

The environmental regulations also influenced the big businesses to adopt practices that are less detrimental to the environment. The predominantly preventive and coercive character of these regulations also stimulated permanent innovations in environmental control. For these businesses, then, the regulations played a relevant role in the adoption of practices that are ecologically less offensive, but these regulations were practically meaningless at the *garimpos* sites.

If any legal norm is to be observed, it must be suitable for the target organizations; government entities responsible for elaborating the norm and executing inspection must be technically prepared for their task; and the necessary social, political, and economic forces must be present to ensure the feasibility of execution. This is anything but the case at the *garimpos* mining sites. The legal norms are inadequate; the entities responsible for application of the norms are poorly equipped for their task; and the social, economic, and political forces are more favourable to maintain the status quo in *garimpo* mining than to transforming this activity into an ecologically less degrading form of production.

Adoption of environmentally more sustainable practices at Brazilian *garimpos* is impeded by

- The existence of a growing contingent of unemployed workers, generated by the ever-worsening Brazilian economic crisis;

- The power of the social groups that flourish and solidify around the *garimpos*;

- The importance of the *garimpeiros* at election time;

- The mobility of these operations and the difficulties this causes for inspection entities; and

- The fact that institutional normalization of these activities would demand extra costs on such items as environmental licences, taxes, and other administrative expenditures.

In this framework, the curtailment of environmental degradation provoked by gold mining in Brazil is fundamentally a question of mitigating the grave social and economic problems that afflict Brazilian society. As well as the creation of job opportunities for the underemployed labour now at the *garimpos*, the following measures are suggested:

1. Improve the structure and preparation of government entities responsible for formulating mineral and environmental policies so that they can more efficiently design and elaborate norms.

2. Develop, improve, and disseminate alternative, nonpolluting technologies for the *garimpos* and small- and medium-scale gold-mining companies.

3. Provide incentives and technical assistance for transformation of some *garimpos* into small- and medium-scale mining companies, structured to meet environmental-protection demands.

4. Adapt environmental legislation to the specific characteristics of *garimpos* and small- and medium-scale gold-mining companies.

5.   Develop and disseminate technologies that substitute less dangerous products for highly toxic inputs.

6.   Provide environmental education programs for *garimpeiros* and for communities near the prospecting areas.

7.   Adopt methods to minimize the occupational health and safety problems in the state and formal-sector mining firms, principally in the underground mines.

CHAPTER 6[1]

ENVIRONMENTAL ISSUES IN BRAZILIAN TIN PRODUCTION

Teresinha Andrade

The purpose of this paper is to summarize the principal results of *Mining and Environment in Brazil: The Case of Tin*, according to the suppositions of MERN (Warhurst 1991) and of the Brazilian team's work plan (Rattner et al. 1991). The essential goal was to determine the spatial and temporal distribution of tin production in Brazil and then map the principal agents and the causes of the environmental problems. Fieldwork included interviews and visits to the more important companies and to the Bom Futuro *garimpo* (artisanal) site. Specialized consultants were contracted to carry out work in the various areas covered by the study.

Organization of cassiterite mining in Brazil

If one is to determine the environmental effects of the tin industry and understand the behaviour of companies and their strategies in the light of a new environmental paradigm, one must know how cassiterite mining is organized in Brazil and find out how it differs from similar activities in other producing countries.

Tables 1, 2, and 3 show world reserves, production, and consumption, respectively, of tin in recent years. It is interesting to note that cassiterite is found mainly in developing countries, whereas it is consumed mostly by industrialized and nonproducing countries. Today, four countries — Brazil, China, Indonesia, and Malaysia — alone produce about 50% of the total, and two — the United States and Japan — consume 40% of the same total. Various factors have been modifying countries' market share and the profile of investments in this industry

---

[1] This paper is part of the Brazilian project coordinated by Henrique Rattner, on gold, tin, and aluminum in Brazil, integrating the overall Mining and Environment Research Network (MERN), which is coordinated by Alyson Warhurst.

Table 1. Principal world tin reserves, principal countries, 1960–91.

	Reserves (1000 t)						
	1960	1969	1974	1980	1985	1990	1991
Australia	48	82	250	350	180	200	200
Bolivia	749	493	990	980	140	140	140
Brazil	—	—	—	130	250	752	1 200
China	—	—	1500	1 500	80 [a]	1 500	1 500
Indonesia	567	559	—	1 550	680	680	680
Malaysia	1 016	610	913	1 200	1 100	1 100	1 100
United Kingdom	48	38	150	260	90	90	90
Zaire	206	157	200	310	20	20	20

Source: Data up to 1974, World Minerals Availability; 1975–2000, SRI (1976); 1981, 1986, 1991, and 1992, DNPM (n.d.).
[a] This marked reduction is not explained by the source of information and continues until 1988, when 400 000* tons is registered.

Table 2. World tin production , principal countries, 1980–91.

	Production (1 000 t)				
	1980	1985	1989	1990	1991
Australia	10.4	9.0	8.0	8.0	8.0
Bolivia	27.3	19.0	14.7	15.0	18.0
Brazil	6.9	26.5	50.2	39.2	32.0
China	22.0	35.0	27.0	27.0	40.0
Indonesia	32.5	20.0	31.5	27.0	27.0
Malaysia	61.4	40.0	32.5	27.0	28.0
United Kingdom	3.0	4.6	5.0	3.0	2.0
Zaire	3.2	3.0	2.0	2.0	2.0

Source: DNPM (n.d.).

at the world level. Such factors include new standards of production and consumption of materials dictated by the consumer countries; the depletion of reserves in some countries and the discovery of new reserves in others; and, chiefly, the decline in the price of tin.

In 1991, Brazil ranked second in world tin production, with 32 000 t, after having been the top producer from 1987 to 1990, when it produced 39 000 t. It also had the second largest ore reserves, or 14.4% of the world total, in 1991. This

Table 3. World tin consumption, principal countries, 1975–91.

	Consumption (1 000 t)				
	1975	1980	1985	1990	1991
Brazil	2	4	4	5	5
France	10	10	7	7	7
Germany [a]	12	15	16	19	22
Japan	28	30	32	34	34
United Kingdom	13	9	8	10	10
United States	44	46	38	38	38

Source: *ITS Bulletin*, WBMS, MMRS.
[a] Does not include former East Germany.

industry's fast growth in Brazil was due to important discoveries in the northern region, in Amazonia, mainly in the 1970s and 1980s, and to government policies offering incentives for the very heavy investments made in that period.

The discovery of a vast mineral province in the Amazon region not only encouraged mining companies to install themselves but also led to the construction of highways connecting the northern region with south-central Brazil and to the introduction of colonization projects. The lack of minimum infrastructure to absorb the wave of migrants in search of new opportunities resulted in the emergence of the region's environmental problems — the devastation of the Amazon rain forest and the degradation of the soil — and aggravation of the social situation, including the displacement of rubber tappers and the local tribes.

Table 4 shows the spatial distribution of cassiterite reserves and tin production in Brazil and points to one of the principal characteristics of tin mining there: 95% of the mining and dressing of cassiterite is done in the Amazon region, whereas most of the tin smelting is done in the more industrialized southeastern region. Of course, this results in widely varying environmental effects.

Another characteristic of cassiterite strikes in Brazil is the predominance of mineralization in secondary material. This means that the ore can be mined by *garimpeiros* (prospectors). In 1990, *garimpeiros* produced 60% of the total mined, which must be considered when one is analyzing the environmental aspects of tin mining. *Garimpo* production of cassiterite in Brazil, which has always been significant, rose considerably when the Bom Futuro mine was discovered in 1987 in the state of Rondônia.

Table 4. Cassiterite production and reserves in Brazil, 1990.

	Production (m³)	Contents (kg)	Tenor (g Sn/m³)	Reserves (m³) Known	Reserves (m³) Estimated
Total	274 519 111	237 769 212		129 048 827	220 842 611
Amazonas	63 519 111	148 621 467		34 749 678	24 536 209
Novo Airão	3 259 739	2 298 115	704	766 560	940 000
Presidente Figueiredo	49 156 737	145 015 905	2 905	19 792 783	2 456 209
Urucará	11 460 902	1 307 447	114	14 190 335	21 140 000
Goiás	18 080 105	26 908 786		32 023 711	161 708 697
Cavalcante	14 013 318	23 869 275	1 703	24 347 419	151 671 181
Ipameri	104 100	623 975	5 993		
Minacu	348 480	204 209	585	368 641	215 136
Monte Alegre de Goiás	231 073	37 433	161	110 825	4 436 792
Nova Roma	2 018 339	932 472	461	5 220 267	4 124 894
Uruaçu	1 364 795	1 241 422	909	1 976 559	1 260 694
Mato Grosso	5 995 643	3 009 812		1 816 000	
Aripuana	5 995 643	3 009 812	501	1 816 000	
Minas Gerais	10 560 795	3 809 232		5 930 109	2 072 797
Aracuai	1 225 035	1 238 510	1 010	2 705 211	

Cassiterita	418 601	5 331	12	16 500	8 500
Coronel Xavier Chaves	442 913	210 850	476	230 000	620 800
Cristiano Otoni	782 401	99 156	126	298 000	
Itinga	2 255 054	474 536	210	2 397 538	1 201 583
Nazareno	8 126	2 730	335		
Ritapolis	4 840 499	1 735 259	358		
São Tiago	588 166	42 860	72	332 860	241942
Pará	35 607 172	19 800 107		11 992 660	1 608 217
Altamira	1 836 955	1 500 246	816		
Itaituba	1 498 557	1 335 214	890		
São Félix do Xingu	32 271 660	16 964 647	525	11 992 660	1 608 217
Paraíba	247 519	473 009		136 034	186 546
Juàzeirinho	247 519	473 009	1 911	136 034	186 546
Rio Grande do Sul	569 680	373 412		128 260	279 000
Encruzilhada do Sul	569 680	373 412	655	128 250	297 000
Rondônia	139 580 819	34 773 387		42 222 385	30 433 145
Ariquemes	36 949 631	19 642 360	531	24 500 629	8 235 345
Porto Velho	102 631 183	15 131 027	147	17 721 756	22 197 800

Source: DNPM (1990).

The outcome in the ensuing years was the following:

- Superproduction of cassiterite, which flooded the international market, causing prices to drop and forcing Brazil to participate as an observer in purchase and sale agreements;

- Corporate reorganization so as to guarantee companies a share of the ore at Bom Futuro, under an agreement with the *garimpeiros*; and

- Disclosure of the environmental destruction caused by tin mining, as much by companies as by the *garimpos*.

Nature and extent of environmental degradation

In the mining and dressing stages, the environmental effects are related to the large quantities of soil and water involved in mining alluvium. For mining, one has to clear land, divert any waterways overlying the cassiterite deposits, build dams, and tear down river banks. Concentration is done in plants coupled to the dredges used in the excavation work. The tailings return to the river or go to tailings ponds, forming mud, which frequently overflows during the rainy season.

The special features of the environmental and socioeconomic problems of cassiterite mining in Brazil as they relate to formal-sector mining may sometimes differ from those that relate to *garimpo* mining, but they should be understood and solved as a whole issue.

Two examples may be used to illustrate differences in ways of producing cassiterite in Brazil and in the environmental aspects:

- Bom Futuro, where *garimpos* cooperatives and organized companies compete for cassiterite in a situation of environmental degradation and where responsibilities for the cost of restoring the environment are not defined; and

- Pitinga, where one company has sole authority to prospect for cassiterite and is directly responsible for restoring the degraded areas.

Environmental effects of mining by a company: the Pitinga mine in Amazonas[2]

Mineração Taboca, of the Paranapanema group, installed and works the Pitinga mine for mining and concentrating cassiterite in the municipality of Presidente Figueiredo, state of Amazonas, about 250 km north of Manaus city.

The Pitinga complex, which includes, besides the mine, a housing estate, complete road infrastructure, and power-generating plants, began operation in 1983. It has already used up investments of about 200 million United States dollars (USD) and, up to 1990, produced foreign currency credits of 915 million USD (Epstein 1992). The accumulated production of metallic tin (until 1990) was 120 000 t, and the annual production represents 15% of world production. This mine is considered the most productive in the world. The venture provides about 2 000 direct jobs and possibly 40 000 indirect jobs. The housing estate, with its population of 5 000, is a complete town, with some infrastructure: treated water, electricity, sanitation, schools, restaurants, medical and hospital facilities, a bank branch, a post office, a telephone exchange, a supermarket, and a business centre.

The 14 areas covered by mining concessions add up to 130 000 ha, but the total area up to now committed to the venture measures about 7 200 ha, that is, 5.5% of the total area permitted.

The Pitinga mine consists of alluvium that has been washed from the primary cassiterite in the bedrock into the valleys. Mining is done, therefore, in the forested meadows. After the forest has been cleared and the river diverted from the mineralized areas, dams and ponds are built. The space is limited, but mining here has the potential to be highly damaging to the environment because the meadows are revolved many times, and it is impossible for them to return to their earlier state. Primary deposits are also mined on the slopes of the Serra da Madeira.

Ever since the Pitinga complex was first installed, Mineração Taboca has tried to reduce the environmental degradation, to do research to provide technical support for environmental-recovery work, and to mitigate these effects. The more important steps have included building dams and dikes for holding tailings and clarifying water, monitoring the quality of the water, replanting mined areas and the slopes of highways, setting up runoff drainage systems, urbanizing and landscaping the operational-support systems and housing estate, mapping degraded

---

[2]   Adapted from Garrido Filha (1992) and Epstein (1992).

areas from satellite photos, and developing environmental-education programs. Important research has been done to determine the adaptability of native and exotic plant species in the region, the effects of interrupting the dormancy of seeds, and the possibility of stocking the mine's tailings ponds with fish; to take forestry inventories; and to define the region's soils.

Before undertaking restoration of the degraded areas, Mineração Taboca prepared a plan and submitted it to the official environmental agencies. Besides taking advantage of knowledge already gained through work in the region, the plan involves the collaboration of research institutes, universities, and specialized companies. The plan covers the area directly affected by the venture for a period of 10 years (1991–2000), although it is expected that mining will continue after that. According to the financial schedule, the cost of the necessary environmental-recovery work is estimated to be 50 million USD (calculated in March 1991).

The initial recovery area — the pilot area — was chosen in 1988 from among Pitinga's most degraded areas. It had been used as a refuse and scrap-metal deposit. At first, it measured 10 ha, but it later grew to 17 ha. The idea was to replant the whole area, using various methods.

Environmental effects of mining by a *garimpo*: the Bom Futuro *garimpo* in Rondônia

The environmental effects of *garimpos* are heightened by the disorderly way in which this mining is done. It revolves larger areas of soil than necessary and results in low output in the ore-dressing stage. Some of the following procedures are very common at *garimpos*:

- Cutting roads in the forest to provide access to mined areas and later intersecting or extending these if necessary to open up another stretch for prospecting or mining;

- Building poorly constructed dams, which often break and contaminate the region's waterways; and

- Pouring tailings into still-mineralized areas because of a lack of knowledge of all the mineral strikes, an action that not only upsets the environment but also means the loss of potential productivity.

At the Bom Futuro *garimpo*, the silting up of the creeks caused by wholesale pouring of tailings has even buried the Vila das Cooperativas, the original *garimpo* nucleus, where in 1987 the headquarters of one of the cooperatives, a school, and a hospital had been installed. The environmental degradation reached such a point that the *garimpo* was closed down by the state government in August 1991.

The environmental issue in Bom Futuro is made worse by the fact that there is no definition of responsibilities for restoring the degraded areas or for preserving the environment. The dispute between the *garimpos* cooperatives and Empresa Brasileira de Estanho S.A. (EBESA, Brazilian tin company) for the mining rights has been delaying measures needed for environmental recovery. After 2 years of legal clashes, in June 1993, EBESA again took over the right to work the mine and to carry out the agreement for working alongside the *garimpeiros*.

Under that agreement, the cooperatives recognize the legitimacy of EBESA's mining rights in Bom Futuro and undertake to sell to the company 80% of the ore they produce. In return, EBESA guarantees that the cooperatives' members may continue to mine the cassiterite, provided they respect the master mining plan and the plan for restoring degraded areas (PRDAs), for which the company is responsible.

Regulatory and institutional issues related to mining and the environment

Starting in the second half of this century, the Brazilian economy was rapidly industrialized and the population became concentrated in urban areas. There were no mechanisms for organizing the space of industrial activity or for providing the necessary controls, so the effects on the physical environment added to the social and economic problems.

The creation in 1973 of the Special Environment Secretariat was the first initiative to institutionalize authority, at the federal level, for preserving the environment. The federal government then introduced systematic rules and published the first decrees governing measures for preventing and controlling industrial pollution.

The National Environmental Policy, instituted by law No. 6938 of 1981, introduced the concept of environmental protection, broadening the scope of the pollution control. As a result, people began to consider the value of managing the environment as a natural resource and of combining economic and social development with the preservation and rational use of the environment (Vianna and Veronese 1992).

The 1988 constitution supported that general approach — analysis of the environmental issue should take into account the social, economic, and institutional aspects. The constitution also determined the decentralization of authority, assigning the legislative and fiscal measures that were to be implemented by the states and municipalities. This legislation applied to all economic activities, including mining. However, some constitutional precepts were instituted for this purpose (these were requirements already stipulated by law). Noteworthy among these were

- The obligation to prepare Estudios de Impacto Ambiental (EIAs, environmental-impact studies) before installing works or carrying on an activity that could cause major environmental degradation;

- The miner's obligation to restore a degraded environment; and

- The imposition of prison sentences or administrative penalties on private individuals or companies whose conduct and activities harm the environment, regardless of the obligation to make good the damage caused.

On one side, the 1988 constitution states that only the central government has authority to legislate on mineral deposits, mines, other mineral resources, and metallurgy. On the other, it stipulates that the states and the federal district have concurrent authority to legislate on the conservation of nature, preservation of soil and natural resources, protection of environment, and pollution control. Consequently, government action regarding prospecting and mining of mineral resources must respect federal environmental legislation and specific state supplemental rules. Miners must ask for environmental permission from their state's environmental agency or from the Instituto Brasileiro do Meio Ambiente e dos Recursos Naturais e Renováveis (IBAMA, Brazilian institute of the environment and renewable natural resources). Such environmental permission first requires an EIA, which must include the following technical activities:

- An environmental diagnosis of the area;

- A description and analysis of the mineral resources and their interactions so that the environmental situation may be defined before the project is installed;

- A definition of measures to lessen adverse effects; and

- A program to monitor the beneficial and adverse effects.

The relatório de impacto ambiental (RIMA, environmental-impact report must state the findings of the EIA and include

- The aims and basis of the project;

- A description of the project and its technological alternatives;

- A synthesis of the diagnosis;

- A description of the probable environmental impacts and how the activities operate;

- A description of the area's future environmental quality;

- A description of the expected effects of measures designed to lessen the adverse impacts;

- A program to monitor the impacts; and

- A recommendation for the most favourable alternative.

When necessary, the government agency will hold a public hearing to obtain information on the project and its environmental impacts and to discuss the RIMA. Some states and territories include this as a requirement in their own constitutions.

Also, when the EIA and RIMA are delivered, the applicant must submit to the appropriate official body a PRDA, as required by the federal constitution. The objective of the plan should be to restore the degraded site according to a pre-existing land-use plan and to achieve environmental stability.

The penalties for violations of environmental rules include

- Fines;

- Loss or restriction of tax incentives and benefits granted by the public authority;

- Loss or suspension of the right to receive official credit financing; and

- Temporary or permanent suspension of mining activities.

Also, the following legal provisions are very important:

- Any mineral prospecting and mining work in an area subject to environmental conservation, as defined in specific regulations, requires permission from the environmental agency that manages that area;

- Any mining work that causes damage to the environment may be temporarily or permanently suspended if so recommended in a report by the appropriate environmental agency; and

- The holder of the prospecting permit, *garimpeiro* permit, mining concession, or mine licence or manifest is answerable for damage caused to the environment.

Furthermore, the federal constitution guarantees *garimpeiros* the right to organize themselves into cooperatives. In certain situations, the government grants these cooperatives a special priority when approving mining authorizations or minerals concessions. The constitution states, however, that the guarantee is conditional on protecting the environment and fostering economic and social aims.

Nevertheless, regulation in accordance with this constitutional precept is controversial because it submits *garimpo* mining to the economic and production logic of organized mining. The *garimpeiros*, even if grouped together in cooperatives, do not have the organizational structure, professional standing, or production flow to enable themselves to meet the legal requirements for preserving and restoring the environment.

Another aspect of Brazil's regulatory regime that affects mining is taxation. Recent changes in the tax system are still stirring up controversy. Particularly controversial are higher taxes and the reduced competitive ability of local producers in an economy that is opening up to foreign competition.

A recent study sponsored by Sindicato Nacional da Indústria de Extração do Estanho (SNIEE, national union of the tin-mining industry), concluded that the tax imposed on tin exports is much higher in Brazil than in Australia, Bolivia, Indonesia, or Malaysia. In Brazil, this tax is 33.35%, whereas the average tax of these other countries is only 11.72%. The study then recommended that to guarantee the competitiveness of Brazilian products in the international market, the export taxes should be reduced to the lowest possible level.

When the Departmento Nacional de Produção Mineral (DNPM, national department of mineral production) analyzed the study, it objected that the main issue is whether tin producers are operating profitability. Even with higher taxes than in other countries, if miners are making a satisfactory profit, there is no reason to reduce taxes. Lower taxes might increase the profit made by the companies but would not increase exports. Furthermore, the DNPM did not want to favour the tin industry by lowering its taxes while maintaining the taxation levels applied to other sectors (DNPM 1992).

This is in essence the regulatory environment in which the mining industry operates. On the whole, although Brazil has no systematic set of rules regulating the sector, mining policy is delineated in the federal constitution and common law (Barreto and Coelho Neto 1993).

Barreto and Coelho Neto's (1993) analysis of Brazil's environmental legislation pointed out some gaps that need to be filled in with supplemental laws until the new mining code is approved. The coding of legislation is always a slow process, requiring political and technical juridical maturity, as the legislation must be perfectly integrated with medium- and long-term sectorial policy. Also, a system of technical rules is still being developed to support and guide the preparation of EIAs and RIMAs and take into account the specific requirements of mining ventures. Without such data, the controlling bodies and law enforcers will be unable to really appraise and estimate the extent of environmental damage and the amounts required to compensate for it.

The mining companies and environmental behaviour

Companies operating in the tin sector

The tin sector represents about 10% of Brazilian mineral production, of which exports account for 8.6% (about 99 million USD in 1991, according to DNPM data). Investments in tin production, involving the work, mine facilities, and dressing technology, reached 45.5 million USD in 1989 — Brazilian production's best year (DNPM 1990).

Production and sales of cassiterite in Brazil are essentially controlled by private groups that have diversified interests in other sectors of the economy besides mining. Table 5 shows the production and sales figures for the companies in 1990; the table also lists those that sell only cassiterite and tin.

Paranapanema Group is the largest local private group engaged in mining. Today, it is the biggest tin-producing company in the world. Its reserves have not yet been entirely measured, especially in the Pitinga area, Amazonas, where it works through Mineraçao Taboca. For tin metallurgy, the group has the company Mamoré Mineraçao e Metalurgia which produced 18 000 t of metallic tin in 1991

Table 5. Brazilian tin production and sales, 1990.

Company	Sn contained in concentrate (t) [a]		Metallic Sn production (t)	Sales to customers (t) [b]	
	Production	Purchase		Domestic	Foreign
Paranapanema	15 691	1 270	16 289	955	15 228
Cesbra	906	291	1 474	996	1 218
Bera	35	260	176	218	120
Best	242	262	481	568	428
Canopus	987	—	985	—	1 025
Mettalurg	78	155	199	38	177
Corumbatai	—	991	998	3 317	50
SNA	—	426	346	245	100
EBESA	2 619	—	—	—	—

Source: SNIEE.
Note: Cesbra, Compañía Estanifera do Brasileira; EBESA, Empresa Brasileira de Estanho S.A.
[a] Refers to tin contained in concentrate, own production and purchase from others.
[b] Refers to domestic- and foreign-market customers.

(nominal capacity 30 000 t/year), or more than 70% of Brazil's production and more than 15% of world production. Paranapanema directly controls the sale of tin to final consumers. It also has a 49.7% interest in EBESA, which was founded to work in the region of Ariquemes, Roraima, specifically at the Bom Futuro *garimpo*.

Mineraçao Canopus, the sector's second largest company, at first belonged to the Rhône–Poulenc group (Rhodia) and now is part of the Silex group, a financial and export conglomerate. The company is not verticalized, which was one of the reasons why it was sold. Because of the new constitutional provisions, it does only prospecting and mining; it sells its production in the form of cassiterite concentrate.

The Grupo Brascan de Recursos Naturais, which derived from the Brazilian Traction, Light & Power Co. Ltd, mines principally through Mineraçao Jacundá (tin mining) in Rondônia and engages in refining and metallurgy through Compañía Estanífera do Brasil, in Volta Redonda, Rio de Janeiro.

The Grupo Best comprises more than 10 mining companies. It has full control of some of them and an equity interest in others and is a verticalized group. It has a foundry in São Paulo, with a capacity of 2 400 t of tin/year (Best Metais e Soldas S.A.) and owns an interest in the foundry of Compañía Industrial Amazonense, in Manaus, whose production is currently suspended (nominal capacity of 3 600 t/year).

The Grupo Brumadinho produces tin through various companies, the most important of which are Compañía Mineraçao São Lourenço (54% Best) and Mineraçao Oriente Novo. Most of these companies' production areas are leased to others. The group works in the mechanical area (making mining, ore-dressing, and ore-concentrating equipment), through Cimaq S.A., and in casting and tin refining, through Bera in São Paulo, whose refining capacity is 5 400 t/year.

EBESA was founded in March 1990. It groups together mining companies producing about 90% of Brazil's tin. Led by the Paranapanema Group, which holds 49.7% of EBESA's shares, they are Cesbra (14.9%), Grupo Best (10.8%), Brumadinho (10.8%), CIF (5.4%), SNA (5.4%), and Impar (3.0%).

According to the companies, EBESA was opened for the following reasons:

- To strategically democratize production control (to avoid a monopoly);

- To control production (to adjust to the market); and

- To guarantee the supply of raw materials to EBESA's members.

In actual fact, EBESA's role has been to negotiate with *garimpeiro* cooperatives so as to control prospecting of cassiterite and avoid more damage to the market, already largely affected by the drop in prices.

Organization of companies

Cassiterite mining in Brazil is all privately organized; there is a strong predominance of local capital represented by the world's largest company in the sector (Paranapanema). This, however, does not represent much in terms of gains in competitiveness or technological mastery in sectors that are strongly oligopolistic at world level, such as the tin sector and the mining industry in the broadest sense.

Furthermore, entrepreneurial activity is extremely verticalized; the same groups control everything from the extraction of ore to even local and foreign tin sales. This is clear from the formation of EBESA, with 75.5% of its shares controlled by the three largest corporate groups.

The tin-mining companies belong to industrial associations in the states where they operate: the Association of Industries of Amazonas State, in the case of Paranapanema, and the Association of Industries of Rondônia State, in the case of those working in Rondônia. All tin-ming companies are members of SNIEE and of Instituto Brasileiro de Mineração (IBRAM, Brazilian institute of mining). Because some have their headquarters in the southeastern region, they are also members of the Federação das Indústrias do Estado do São Paulo (federation of industries of the state of São Paulo).

Organized in this way, the companies are asking for definitions of the right to mine and demanding the mechanisms to defend such rights. They are also criticizing environmental controls: very often, requirements overlap, with authorities at different levels acting without coordination. They are also denouncing the chaotic assortment of taxes imposed on the mining sector.

IBRAM organizes and represents the mining companies on the environment through the Technical Environment Committee, set up in 1983. This committee has been giving guidance and instruction to help combine mining activities with protecting the environment, involving its members, the government, and the community. Through IBRAM, the mining companies have submitted studies to the federal authorities, showing that the tax burden borne by Brazil's mineral sector is one of the world's heaviest, making it difficult to increase productive investments. The mining companies are also complaining about the lack of specific credit lines at beneficial interest rates for equipment to prevent and remedy pollution, a situation that is common in other countries where mining is traditional (Epstein 1992). In April 1989, IBRAM signed a technical-cooperation agreement with IBAMA to carry out studies and suggest solutions to environmental problems related to mining activities. Contributions have been made on disclosure and clarification of the application of Resolution No. 1/86 of the Brazilian Comisión Nacional del Medio Ambiente (CONAMA, national commission of the environment) to EIAs for mining ventures and on the disclosure and implementation of methods for restoring mined areas, among other initiatives (Viana and Veronese 1992).

Once the need to reconcile mining with environmental protection was made clear, IBRAM began to play its part by disclosing and encouraging discussion of issues, through congresses, specific workshops, courses, and various papers on ideas and more modern mining methods.

Environmental policy and corporate investments in technology

The mining companies see environmental control as part of industrial operation and have environmental departments. Work done by the larger companies includes extraction planning, disposal of tailings, control of the dressing phase, and recovery of mined areas, although the work has been mostly restricted to prompt action aimed at controlling pollution.

Because mining is openly aggressive toward the environment, it is natural that society constantly endeavours to limits its effects. But in Brazil, most of the mining is done in the northern region, where the power of even government agencies is restricted because of the area's immense size and low population density. A power struggle is going on between the mining industry and society about the rules to protect the environment. Evidently, mining companies prefer lenient environmental rules to avoid overloading production costs, and they take every opportunity to state their case. The state should, therefore, act not only to maintain control of this situation but also to develop a feasible approach to dealing with the social, economic, and institutional aspects of environmental issues, as determined by the 1988 constitution and recent complementary legislation.

Coordinated measures for preserving the environment are still lacking in the tin sector. As mentioned earlier, investments in this area are almost entirely confined to restoring degraded areas and to preparing studies and research on local ecosystems so that the extent of the damage can be monitored. Although technological innovations stimulated by the pressing environmental problems may eventually reduce both the production and the environmental costs, cassiterite-mining companies are not yet making significant investments in this area.

Some advances have been made in methods of implementing technologies already known in cassiterite mining, such as layout modifications. New basic flow layouts were introduced that are more compact and versatile, reducing the need to clear land for building approach roads. Similarly, the substitution of equipment driven by firewood for electrically driven equipment, besides minimizing costs, reduces environmental impacts (Epstein 1992).

Cassiterite mining will increasingly have to use cut-and-fill methods, not only to reduce environmental impacts but also to minimize transport costs and tailings ponds. Gravimetric methods of concentration will continue to prevail, although gradually new, more compact and versatile equipment should be introduced to provide more agility in joint work on the mining fronts.

Labour profile and the social and regional problems of formal and *garimpo* cassiterite mining

According to the Fundação Instituto Braseleiro de Geografia e Estatistica (IBGE, Brazilian institute of geography and statistics) 1985 industrial census, workers engaged in formal-sector mining had almost 2% of the total industrial jobs in Brazil. If industrial processing (metallurgical) is added in, this share would be 18.5% (in 1985). Tables 6 and 7 show the distribution of such labour in 1979–89 according to the standard of education and percentage used by the tin industry.

Table 6. Total labour (*n*) in the mining sector and percentage of labour in the tin sector, 1979–89.

Education or activity	1979	1981	1983	1985	1986	1987	1988	1989
Graduate	1 689	1 959	1 997	2 481	2 417	2 355	2 531	2 692
%	4.4	5.6	6.1	8.2	7.6	4.9	5.6	4.8
Intermediate	1 886	2 459	2 557	3 354	3 561	3 888	4 022	4 330
%	5.6	2.5	6.3	5.6	7.8	3.8	3.3	3.3
Labour	59 978	63 926	68 596	76 766	78 535	77 631	79 148	8 4319
%	4.9	4.4	5.5	7.7	6.8	5.7	5.2	4.1
Administration	6 849	8 586	8 138	10 591	9 651	10 027	12 004	11 493
%	4.0	4.4	6.3	18.2	11.95	8.9	6.7	4.3
Total	70 402	76 930	81 288	93 192	94 164	93 901	97 705	102 834
%	4.8	4.4	5.6	8.9	7.3	6.0	5.3	4.1

Source: DNPM (n.d.).
Note: Percentages refer to labour used per period in the tin sector as a percentage of the total used in mining.

Table 7. Labour used in mining cassiterite, 1979–89.

Education or activity	1979	1981	1983	1985	1986	1987	1988	1989
Graduate	74	109	121	204	184	115	142	129
% var.	—	47.3	11.0	68.6	−9.8	−37,5	23.4	−8.5
Intermediate	105	61	161	187	277	147	133	142
% var.	—	−41.9	63.9	16.1	48.1	-46.9	−9.5	6.8
Labour	2 918	2 838	3 775	5 945	5 302	4 445	4 136	3 436
% var.	—	-2.7	52.3	57.5	−10.8	−16.2	7.0	−16.9
Administration	276	379	515	1 932	1 153	896	800	493
% var.	—	37.3	35.9	75.1	−40.0	−22.2	−10.7	−61.6
Total	3 373	3 387	4 572	8 268	6 917	5 603	5 211	4 200
% var.	—	0.4	35.0	80.8	−16.0	−19.0	−7.0	−19.0

Source: DNPM (n.d.).
Note: % var., annual percentage variation, using as a basis (100%) the data of the preceding period.

A decline is noted in labour employed in mining cassiterite from 1985 and in its participation in the mineral sector as a whole, as a result of the massive entry into the production market of the Bom Futuro *garimpo* production and the drop in international prices.

The formal mining sector does not attract much labour, except in large projects, and no such projects are at present engaged in mining cassiterite. In the mining companies, workers are paid wages. Most of the specialized technicians, particularly university graduates, come from the south-central Brazil and receive extra benefits to set themselves up in remote parts of in Amazonia, such as Pitinga, Amazonas; Massangana and Santa Bárbara, in Rondônia; and São Pedro Iriri, Bom Jardim, and São Raimundo, in Pará. Unskilled labour is usually recruited from the local population. Although the companies have plans for restoring degraded areas and are working out environmental-management methods, no significant changes in the number or profile of labour used have been detected in tin mining.

In the mechanized *garimpo*, the mining system is work intensive. Although heavy equipment is used, labour is very important at the *garimpos*, which in 1988 employed about 20 000 *garimpeiros*. To define the profile of labour used in the *garimpos*, first it is necessary to distinguish the worker *garimpeiro* (*requeiro*) from the entrepreneur *garimpeiro* (*produtor*). The latter owns the equipment used in the mechanized *garimpo*, such as tractors and power shovels, and monopolizes ore sales. The entrepreneur *garimpeiros* usually come from southern Brazil and almost always are lumberjacks who accumulated this activity with the *garimpo*. The worker *garimpeiros* are usually from the northeast, particularly the state of Maranhão, or they come in from neighbouring farming areas between harvests. In Bom Futuro, the *garimpeiros*, despite being organized into cooperatives, give no thought to contracting qualified professionals to prepare and implement a mining plan. They do, however, contract lawyers and expediters to help them understand and respect the laws, as a requisite to waging the war they have declared on the companies.

Also, it is necessary to consider the seasonal nature of the *garimpo*, where the work is mostly done by peasants who usually return to their farms when there is less work in the *garimpo* (Garrido Filha 1983). Hence, although the land in Brazil is undeniably concentrated because of the lack of agrarian reform, the migration of peasants to the *garimpos* of the Amazon region is more an effect of the hard conditions of rural activity and generally of the illusory search for better living conditions by people unprotected by the prevailing economic model.

Usually, the mine is in a remote rural zone, often with no other economic activities. What does happen, and not rarely, is that settlers from neighbouring farming projects invade the areas applied for by a mining company (Garrido Filha 1992). As mentioned, the company builds its village, with a supply of water, electricity, and basic sanitation. Schools are opened, as well as medical clinics or even small hospitals. A supermarket or stores sell essential retail goods, and everyone has adequate lodgings. *Garimpeiros*, especially when the *garimpos* are large, such as Bom Futuro, install nuclei, without any concern for organizing space and basic sanitation.

Government policies and company strategies

The action of the state and of society

Environmental management in Brazil has relied almost entirely on command-and-control mechanisms (Reis and Motta 1992). These mechanisms, according to law No. 6 938, of 31 August 1991, which governs the National Environmental Policy, can be divided into four categories:

- Environmental standards (quality and emissions);

- Control of land use (zoning and protected areas);

- Licencing (EIAs and RIMAs); and

- Penalties (fines, compensation, etc.).

Although various systems have been organized at federal and state levels to implement the various measures assigned by the constitution, environmental action has been jeopardized by the shortage of funds and human resources and by poor integration between and inside government levels and bodies.

The implementation of environmental programs depend principally on funds from abroad. Table 8 clearly shows the scanty funds available for the development of science and technology for the environment or for training personnel. Central-government budget figures, which are far from being the actual expenditures on implementation, amount to not even 1% of the total federal funds collected in the same year.

Table 8. Principal federal funds for C&T in environment and mining, 1991.

Public institutions and programs	×1 000 BRE	×1 000 USD
On national scale	111 916.008	273.467
IBAMA	88 276.748	215.704
Environment Secretariat	11 705.863	28.603
PADCT (Geoscience and Mining Technology)	4 569.049	11.164
Regional Development Secretariat	4 374.378	10.689
CETEM	1 301.205	3.180
PADCT (Environmental Sciences)	1 195.586	2.921
Development of environmental technology	493.179	1.205

Source: *Central Government Budget* (1991).
Note: BRE, cruzeiros; CETEM, Centro de Technologia Mineral (centre for mineral technology); IBAMA, Instituto Brasileiro do Meio Ambiente e dos Recursos Naturais e Renováveis (Brazilian institute of the environment and renewable natural resources); PADCT, Programa de Apoio ao Desenvolvimento Cientifico e Tecnológico; USD, United States dollars.

Society is sympathetic to the environmental issues and has been stepping up its pressure on the government to see environmental management expanded from purely corrective measures to support for action on long-term sustainable development in economic, social, and ecological terms. To attain this standard of environmental management, some adjustments must be made in the structure of the environmental agencies to offer stable conditions for achieving medium- and long-term objectives and implementing programs. Furthermore, it is fundamental that society be mobilized, to really share such management (Cerqueira 1992). There is still a long way to go. Much of the population has to contend with acute environmental problems, but the economic and social problems of subsistence tend to exclude ecological issues from the principal anxieties of most Brazilians (Olsen 1992).

Although the state's and society's demands are not yet backed up effectively, the mining companies have realized that even to market to society, they have to be constantly fighting against environmental degradation. All the same, the companies have had difficulties putting this policy into practice. Various problems have arisen with the communities and fines have been imposed by environmental agencies when mining has been done on Indian reserves or in areas designated for permanent preservation, not to mention the penalties for failure to carry out the companies' plans for restoring degraded areas, as widely disclosed in the press.

The sector's investment schedule

The mining industry in Brazil showed its best performance in the 1980s. The sector's weight in industry as a whole, because of the value of industrial processing, was 5% higher in 1985 than in 1980, according to the IBGE's industrial census.

Mining today ranks in fourth place among the 22 industrial sectors, behind chemicals, metallurgy, and food products. These three sectors are closely linked to mining, whether through their inputs or through derived products, such as food packaging. Hence, the most dynamic sectors of the Brazilian economy are providing opportunities for mining companies to diversify their business.

If, in the short term, Brazil's tin sector, together with the government, intends to increase its influence over tin production and exports, in the medium and long term, the tin companies will tend to diversify into other sectors or to verticalize to safeguard against the normal fluctuations in primary ore markets.

As an immediate consequence of the situation in the international tin market, companies were announcing some production cuts for 1993, although they will have had few effects on the estimated domestic production of 25 000 t for 1993 (26 000 t was produced in 1992). The decline in production of some mining companies should be offset by the increased production at Bom Futuro. In addition to acquiring concentrate furnished by the *garimpo*, EBESA intends to increase its own participation. By 1993, it has already invested 4 million USD in mining equipment and infrastructure, and it intended to invest 2.5 million USD more by the end of the year (*Gazeta Mercantil* 1993).

Final comments

The Brazilian tin industry has been going through some stressful times: the persistently low price of tin in the foreign market; market displacement of primary tin by alternative materials and recycled tin; China's entry into the market as a large producer; and higher production costs as a result of having to implement PRDAs.

As well as making production cuts in response to market conditions, the sector has reorganized itself. One example of this was the founding of EBESA to deal with the production of the *garimpeiros*, who had organized themselves into cooperatives in Bom Futuro. One should not disregard aspects such as this when considering the environmental issue of cassiterite production in Brazil.

Depending on the stability of the tin market, the companies are prepared to make investments in technology, infrastructure, human resources, and environmental protection. By and large, these companies are already aware of the need for environmental controls and for restoring areas degraded by mining. In many cases, though, the companies are taking measures related to the environment only because there are legal requirements to do so, although society's increased demands are already having some effects.

Both the state and the companies are still finding it difficult to move beyond simply remedying specific effects on the environment to integrating the social, political, economic, and technological aspects of environmental management, which are inextricably related. The need for this integration is strongly felt in light of the fact that cassiterite production is divided between the strongly oligopolistic corporate sector and *garimpeiros* and takes place in the Amazon rain forest. Steps companies take to restore degraded areas still require medium- and long-term assessments, regarding not only the final effects of efforts to restore the environment but also the mechanisms and methods used.

The most serious environmental impact of mining, which the PRDAs and other isolated measures have not managed to solve, is the silting up of rivers and creeks. This degradation modifies forever the profile of animal and plant life, destroys gene banks, alters the soil structure, introduces pests and diseases, and creates an irrecoverable ecological loss.

The lack of technological developments specifically for mining cassiterite is also felt. The PRDAs are aimed at mitigating the problems caused by currently used technologies, but the companies should also be investing in new technologies. The expectation that the development and adoption of clean technologies might reduce operating costs and increase efficiency is still insufficient to motivate companies to make such investments. Rather, the companies are motivated by the profit margin that the current market offers.

Because the environmental impacts of cassiterite mining in Brazil are complex and interrelated and their full spatial and temporal effects are still unknown, dealing with the environmental issue requires government participation. In addition to legislated control measures, broader regional planning and a democratic set of government policies are necessary.

CHAPTER 7

ENVIRONMENTAL MANAGEMENT IN THE BAUXITE, ALUMINA, AND ALUMINUM INDUSTRY IN BRAZIL

Liliana Acero

This paper briefly presents the main results of a study of the bauxite, alumina, and aluminum industry in Brazil, conducted to investigate hypotheses of the Science Policy Research Unit network (Warhurst 1991a) and the Brazilian team within it (Rattner et al. 1991). Examining recent international trends, the study documented how that industrial sector operates in the economy (Acero 1993).

The main objectives were to

- Document specific environment-management practices of various companies operating in Brazil;

- Relate these practices to recent governmental and societal laws and regulations on environmental controls and planning;

- Measure or illustrate possible industrial changes in environmental management and technical solutions resulting from legislation or the pressure of different social sectors; and

- Document community-level environmental impacts of current practices and the firms' responses, where possible.

A word on environmental costs

The basic rationale for environmental costs at the micro- and macrolevels has been widely debated in the past 10 years (see, for example, Hurrell and Kingsbury 1987; Adams 1990; Nappi 1990). Alternatively, it is argued that in developing countries, a level of environmental degradation is an unavoidable condition of economic growth that in the short run diminishes the costs of development; and that unless some environmental degradation is allowed, then enterprises tend to externalize the costs of environmental controls and antipollution measures, costs that are ultimately borne by the public in developing countries. More technically speaking, there has also been some concern about how to discriminate between the environmental and capital costs of any economic activity.

Although in this chapter I do not discuss these issues in depth, one should be aware of the magnitude of the academic and policy problems, which involve changes in the ways management and general administration collect data on economic and financial investments and costs, especially within developing countries. This issue exceeds the scope of the present study. However, I do take a practical approach to the issue here.

First, I tied to collect and analyze the fragmentary type of information that firms gather during their activities. Second, I analyzed, with the interviewees, the main parameters that firms use. Third, I organized the data so that I could do an interfirm comparison. Fourth, in some cases, I requested that similar data be gathered by firms that had not done this before. However, the information should be considered indicative of an order of magnitude, rather than a comparison of practices between establishments.

The firms' costs for their activities tended to be differentiated into two major categories:

- *Operating costs* — These usually include all running costs attributed to the environmental department, such as wages of permanent and subcontracted personnel, material replacement, and input and capital costs of new laboratory materials, and amortization rates of existing capital; and

- *Investment costs* — These involve costs of major environmentally oriented investments, such as specific projects, like the deposition of tailings and red-mud slurry, or major capital investments for antipollution purposes, such as the closure of prebaked cells.

In both cases, similar types of activities were listed for each firm. However, the results should be treated with caution: they are not to be regarded as absolute numbers of real performance. This is for a variety of reasons, some of which relate to the specific characteristics of the production processes for bauxite, alumina, and aluminum. First, some of the machines normally used in the production process embody antipollution technology, such as dry-scrubbing, that in turn is material or input saving. This is not commonly included either in operating or in investment costs. Second, minor technical adaptations considered more environmentally sound but undertaken for reasons of efficiency are also excluded from the parameters; an example of this would be changes in the feeding devices of prebaked cells. So, in some ways, the results underestimate activities immediately related to the environment.

Despite these limitations, these data were relevant to the study because this information provides a reference point for drawing comparisons and because, based on this type of information, enterprises carry out their negotiations with local policymakers and environmental bodies. Ultimately, these data also serve to illustrate the problems with making measurements of this kind in an industry that combines input and materials savings with environmentally sound technical solutions. In the next section this point will be discussed in detail.

Nature and extent of environmental degradation

The three phases of this industry — bauxite mining and exploration, refining, and smelting — can each have many effects that are potentially hazardous to the environment. But before addressing these concerns, I will briefly consider the production process, as described by Warhurst (1991a, p. 23):

> Bauxite is the raw material from which alumina is extracted. Bauxite occurs in a small number of high-grade deposits mainly within a tropical zone extending up to 20 degrees north and south of the equator. The technology of alumina production varies with type of bauxite ores which fall into three groups: Monohydrate bauxite is generally found in boehmite in Europe and northern Asia and is processed using a European version of the American Bayer process (described below); Trihydrate bauxite is found as gibbsite in Surinam, Guyana, Guinea, Ghana and Australia and is processed by the American Bayer process. The Jamaican type of bauxite is characterized by a mixture of gibbsite and boehmite and is found in Jamaica, Haiti and the Dominican Republic. It is processed using a method which combines elements of the European and American Bayer process.
>
> The basic Bayer process involves the following operations ... . Dried, ground bauxite is mixed in a large digester vessel with caustic soda which dissolves the aluminum oxide under strong pressure. Impurities, such as iron oxide and silica, are filtrated out in their solid state. Then the sodium aluminate liquor is seeded with hydrated alumina crystals and part of the solution combines with the "seeds" to form alumina hydrate crystals. These are then calcined in long rotary kilns under high temperatures. This leaves calcinated aluminum as a white powder which is then ready

for transformation into aluminum metal. On average, 2.25 tonnes of bauxite generates one tonne of powder.

Finally, aluminum is separated from its oxide by the highly energy-intensive Hall–Heroult process. The process takes place in carbon-lined reduction cells. First, alumina is dissolved in a molten salt called cryolite to which aluminum fluoride is added continuously to maintain the required density, conductibility and viscosity. Second, a carbon anode is lowered into the solution causing a continuous electric current to pass through the mixture to the carbon cell lining, which acts as the cathode. This causes the dissolved alumina to separate out into aluminum metal and oxygen, and since the former is heavy it is attracted by the cathode to the bottom of the pot, while the oxygen settles on the carbon anode to form carbon dioxide. The molten aluminum in the pot is syphoned into crucibles and transferred to alloying furnaces to make alloys. Finally, the metal is cast in an ingot mould.

The potential hazardous impacts of these processes can be divided into those strictly affecting the physical and natural environment and those of a socio-economic nature (UNEP 1984).

Mining

In ore mining, the environmental effects are highly site specific, but, on the whole, the main impact comes from clearing the vegetation in bauxite mining, exploration, and mine development. A secondary effect is that brought about by the dumping of wastes or the inadequate management of tailings. This can degrade the habitat of local flora and fauna and make future land use difficult for reforestation, agriculture, or cattle breeding.

Mining can also affect local water and air quality. For example, with the removal of overburden, runoff can become contaminated, more acidic, and more turbid. Erosion within the mined areas can be rapid if the soil is not recovered and reforested. The removal of vegetation can then bring about loss of flora and fauna, destruction of wildlife habitats, a possible spread of plant disease, increased soil erosion, changes in weather conditions, dust, and a possible need for runoff water treatment. In open-cast or surface mining, the areas cleared of vegetation may disrupt the landscape and produce a negative visual impact.

Beyond this, fugitive dust and noise from heavy machinery and explosives may be a source of disruption in the external environment for nearby communities and a health hazard in the working environment.

Adverse socioeconomic impacts depend largely on the proximity of the mines to established communities. Impacts can include breakup of cultural traditions, lifestyles, and kinship groups; substantive changes in agricultural crops, techniques, and marketing resulting from weather and soil disruptions; and lack of infrastructure, other employment opportunities, housing, and educational and recreational facilities for the personnel.

Refining

In refining or alumina production, the environmental impact largely depends on the composition and quality of the ore and the processes used to extract bauxite. The principal hazards arise from the disposal or storage of bauxite-residue slurry (red mud), which is the main alkaline effluent from the alumina plants. This is either disposed of on land or discharged (dewatered or untreated) into sealed or unsealed artificial or natural areas or into the deep sea. Liquid and solid phases in slurry have the following potential effects (UNEP 1984):

- Seepage of the alkaline liquid into groundwater, which might contaminate industrial, domestic, and agricultural water supplies;

- Spillage from damaged pipelines or from retaining-dyke failure;

- Reduction in the availability of arable land;

- Dust pollution in arid regions; and

- Aesthetic impacts.

Airborne pollutants (dust and noxious chemicals) are another kind of hazard from stockpiles, mills, and calcination operations. The air pollutants are bauxite, lime, and alumina dust, SO_2, NO_2, dust from low-grade bauxite, and suspended vanadium pentoxide. The quality of SO_2 pollution depends on its concentration in fuel oil, its specific form of consumption, and the ways power is supplied to the plant. Gas emissions not collected or insufficiently collected, especially in the case of SO_2, can contaminate the workplace and the general environment and, in reacting with water, can produce acid rain.

Spills can occur at different stages in the refining process. Acidic drainage, if untreated, can damage local flora, fauna, and even human beings. The work environment can be extremely hazardous where people handle corrosive chemicals, such as caustic soda and acids, or where there are reverberation, fumes, dust, and certain toxic chemicals.

Aluminum production

In aluminum production, the main environmental problem is air pollution caused by fluoride emissions in the smelting process. These can have a strong effect on workers' health inside the plants (excessive intake can cause fluorosis and skeletal disorders), as well as on the surrounding flora and fauna. As molten cryolite is used in the electrolysis of alumina, the fumes emitted from the cell have gaseous and particulate fluorides. Gaseous fluorides are also contained in the exhaust gas of the anode baking furnace and are emitted if unscrubbed. They can also be generated at a lower degree from the cast-house furnaces. Other fumes emitted by some types of cells are tar fumes (that contain suspected carcinogens) and SO_2 (when petroleum coke with sulfur is used for the anodes); SO_2 is also produced by anode-baking and cast-house furnaces, especially in plants using thermal power. At these last two stages, nitrogen oxides are also emitted. Dust is found at different stages in the production of aluminum.

Other problems are polluted water, solid waste, noise, and heat. Water pollutants are stronger when a wet-scrubbing system is used to clean fumes emitted in the cells. The water contains fluorides and suspended solids, such as alumina and carbon, that need to be treated before discharge. The main solid-waste problem is that of spent potliners, as cells have to be relined every 4 or 5 years. The linings can leach fluorides and cyanides to surface or groundwater when they are stored in the open air and in pits. Cast-house furnaces generate drosses that can produce fugitive-dust losses or gases that can evolve into ammonia if wetted. Noise and reverberation in most smelting stages and heat in the potline rooms tend to be at very high levels and insufficiently controlled in Brazil. This can affect workers and inhabitants of local communities.

Another indirect hazard to the environment is due to the intensive consumption of electricity in the aluminum-smelting industry. These plants must be located near cheap sources of electricity, mainly hydroelectric dams. The construction and the operation of these dams also pose a potential threat to the environment. Large areas have to be depopulated and flooded for dam building, changing the ecosystem of a whole area. This has potential negative effects on flora and fauna and even jeopardizes human health in the area.

Economic restructuring

Several factors are contributing to the reduction in the growth rates in aluminum consumption worldwide. The growth of industrial production is slower than that the growth of the service sector in the developed world. The substitution process whereby aluminum gained market shares from other materials has slowed down.

The only new and substantive market for aluminum has been that for aluminum cans (which have replaced tin ones). This market is still expanding, especially in the developing world. In the rest of its final uses, aluminum has maintained its quotas; however, aluminum recycling has increased significantly in the last decade, further reducing the potential expansion of aluminum consumption.

Another reason for restructuring in the industry has been the increasing participation of developing countries, especially through government enterprises in different phases of the industry. Developing countries have tried to make better use of their comparative advantages in natural resources and energy, usually by raising the prices of exported raw materials and obtaining a higher elaboration in mineral production (products with value added). In this sense, the new strategy adopted by large world producers has been either to increase the value added and quality of their products or to explore new types of activities, such as developing new aluminum products (for example, Kaiser Aluminum Corporation) or using more modern materials (for example, Alcoa Alumínio S.A.).

An additional factor has been aluminum's appearance on the London Metal Exchange (LME), which took away the major companies' control of overpricing, removing one of the industry's most important entry barriers.

In 1990, Brazil had an installed production capacity for primary aluminum of 1.132×10^6 t. Bauxite production accounted for 9.875×10^6 t, of which 55.3% was exported in 1990; alumina was of the order of 1.654×10^6 t; and aluminum reached 930 000 t. The domestic price for primary aluminum, as fixed by LME, was internationally competitive: 1 738.69 United States dollars (USD)/t in 1990. This was partly due to the subsidized energy tariffs charged to most of the large aluminum complexes. These tariffs were 25.61 USD/MW·h for high-voltage transmission to ALUMAR and Alumínio Brasileiro S.A. – Alumina do Norte do Brasil (ALBRAS-ALUNORTE), for example, and 27.48 USD/MW·h for low-voltage transmission to firms in the southeastern region (in São Paulo and Minas Gerais). Most employment is in the primary-aluminum integrated firms; in 1990, for instance, 40% of the 66 780 people working in the alumina and aluminum sector worked at such firms. Meanwhile, 1 543 people were employed in bauxite firms that year.

In 1980–90, Brazil had the second largest growth rate in production (257%) among the largest primary-aluminum producers. It was only outweighed by Australia (307% growth). Since 1989 it has become the fifth largest producer in the world and holds the third largest bauxite reserve (10.2% of the world's 22.7×10^9 t). Almost 60% of production was exported in that year. Production continued to grow until the present, with export rates substantially outweighing expansion for the internal market. In 1990, sales in the industry represented 0.8% of the gross domestic product. However, Brazil has one of the lowest per capita consumption rates among the main primary-metal producers (2.7 kg per person in 1989), but it is an average level when compared with other industrializing countries in Central and South America.

The next section will explore the legislation that directly or indirectly affects activities carried out in this sector in Brazil. It will emphasize the implementation of the environmental-control laws since the new Constitution that have had the most recent effect on the industry.

Regulatory and institutional mechanisms

Four types of regulatory mechanisms, established in the last decade or so, apply directly or indirectly to activities carried out in the bauxite, alumina, and aluminum sector:

- The constitutional norms of 1988 relating to the mineral sector;

- Legislation on the exploration and use of mineral resources;

- Environmental legislation on mining and its industrial transformation; and

- Tax laws.

Currently, there is no new mining code functioning (the last one being from 1967). Actual practices are ruled by a set of independent laws, based on the spirit of the 1988 Constitution and passed after it. However, a law project to form a mining code was presented to the national Congress in 1991. The main innovations within the Constitution are the following:

- The republic has sovereignty rights over mineral resources (including those of the subsoil), resources of the archaeological and prehistoric sites, and natural resources of the continental platform. This allows the republic to dispose of these resources as their sole owner and makes the republic obligated to protect them. Dual property rights, over land and subsoil, plus sovereignty rights of the republic over minerals shape the spirit of the new mineral rights and public policy, legitimizing the republic's sovereignty. However, the Constitution allows the owners of mining concessions to have ownership rights to the products they extract.

• The republic may decentralize administrative functions pertaining to natural-resources management. The Constitution establishes that only the republic can pass legislation governing mineral resources, not the states (as had been the case in the former Constitutions). It admits that new laws might be passed allowing the states to legislate on limited and specific questions. In turn, the latter can register and monitor mining concessions, as can municipalities. In future, these tasks will be distributed, depending on technical and administrative competence, between them and ruled by a mining code.

• All levels can pass legislation pertaining to the defence of natural resources, the protection of the environment, and protection controls. If the laws are contradictory, the republic has to fix the more general criteria.

• The republic has the right to limit the form of participation of foreign capital in mining.

• By law, as of April 1991, the National Defence Council (CDN) has authority over the use of national-security areas (especially frontier regions) and the preservation and exploration of natural resources of any type. The Departmento Nacional de Produção Mineral (DNPM, national department of mineral production) verifies whether these demands are fulfilled, because the attributes of the CDN are still formulated in very generic terms. Two of the main new norms are that 66.6% of employees in the companies operating in such areas and the majority of management must be Brazilian citizens.

• The states and municipalities are to receive compensation for the results of natural and mining-resource exploration. This is tied in with tax laws.

• The republic imposes a new system of taxation for the mineral sector and revokes the Imposto Único sobre Mineral (unified tax on minerals), a unique tax.

The forms of mining concessions and their life spans have also changed. Durations of mining-research permits have been shortened. Some entrepreneurial groups have been fighting for a lifetime research concession, a mode of operation that would rule out contractual agreements and is thus internationally outmoded. Moreover, research and exploration authorizations have been restricted to Brazilian firms with national capital or Brazilian individuals. Thus, foreign capital is only allowed to have a minority participation in the firms, whose effective control and capital are to belong to local residents. However, subsequent regulations, known as the "transitory dispositions," ameliorate these limitations by allowing 4 years for companies to adapt to new restrictions and by virtually eliminating the restrictions for those firms that in that time locally promote the industrialization of their mining and beneficiation activities. These firms can receive research and extraction concessions if the minerals are intended for use in local industrial processing. Authorizations can be canceled if they were not initiated in the terms established. The Constitution makes a clear division between the rights of the owners of land and the rights to the subsoil. But it is guaranteed that the owner of the land receives benefits for the results of mining. Finally, mining research and extraction on lands occupied by native populations require Congressional approval; mining titles are suspended in those lands because Congress has still not passed ordinary laws concerning this.

Environmental legislation was virtually nonexistent in previous Constitutions, but the 1988 Constitution devotes a whole chapter to it. The *Mining Code* of 1967 only mentioned generically that the owner of the mining concession should avoid air and water pollution resulting from its works. But already in 1981 the first National Policy for the Environment was passed as a law (law No. 6, 1981), and it established the Systema Nacional de Meio Ambiente (national system for the environment). These previous laws were taken up and given constitutional status in 1988. Four main aspects are considered in the Constitution:

- The firm's obligation to have an environmental-impact study (EIS) before setting up any activity with a potentially hazardous impact on the environment;

- The firm's obligation to rehabilitate degraded environments;

- The establishment of penal and administrative sanctions for the degraders independently of their obligations concerning rehabilitation; and

- The ruling of responsibilities for damage caused to the environment.

Mineral extraction is subject to laws passed by the Republic and by the states and must comply with laws passed by both. Licencing depends on the state or on the

Instituto Brasileiro do Meio Ambiente e dos Recursos Naturais e Renováveis (IBAMA, Brazilian institute of the environment and renewable natural resources), when appropriate. Two resolutions of the Comisión Nacional del Medio Ambiente (National Commission of the Environment), passed in 1989 and 1990, deal specifically with environmental licencing in mining.

For research, the companies require an operational licence, which includes an evaluation of its environmental impacts and an analysis of the measures to be adopted to solve them. For extraction, three types of licences are required: the preoperational licence, which also includes the EIS; the installation licence, which requires a detailed plan for environmental controls; and finally, for actually operating, registration at DNPM (if the ministry approves the operational licence).

The EIS should cover the following technical activities:

- An environmental diagnosis of the area;

- A baseline analysis and description of the environmental resources and their interactions (before the project);

- The definition of suitable measures against negative impacts; and

- A program to monitor the positive and negative environmental impacts.

A second report, the relatório de impacto ambiental (RIMA, environmental-impact report), is accessible to the public. It must

- Discuss the conclusions of the EIS, mainly to outline the aims of the project;

- Describe the technological alternatives;

- Provide a synthesis of the area diagnosis, including a description of possible environmental impacts;

- Characterize the future atmospheric quality of the area;

- Estimate the possible results of the proposed impact controls;

- Establish a detailed monitoring program;

• Recommend the most favourable control alternatives; and

• Provide illustrations (maps and graphics).

The EIS and the RIMA should both include a program for post-operation land recovery and rehabilitation. Both documents have to be developed by a multi-disciplinary team, which must be authorized to carry out this type of work.

A further environmental-policy tool provided for by the 1988 Constitution is public audiences. The public environmental bodies set up these public audiences, at which they present, for discussion, information on the project's probable environmental impacts. This discussion is compulsory in some states, such as those in the Amazon region.

A number of sanctions have also been established for enterprises that fail to fulfil the above requirements: fines, the loss or restriction of fiscal incentives or benefits, the loss or suspension of the firm's participation in government loans, or the temporary or definite suspension of its activities. New developments in environmental laws (a preproject formulated by a commission assessing the Executive and sent to Congress) suggested that two important amendments to the Constitution would be included in its revision, foreseen for 1993. First, all licencing concessions for mineral activities should be concentrated at IBAMA, thus disengaging the authority of the states to rule over and administer them. Second, licencing evaluation should take into account the economic need for mineral extraction. These two clauses, as well as one on *garimpos* that does not relate directly to our research, tend to make procedures more bureaucratic and to centralize decision-making. This contradicts the original spirit of the constitutional text, although it means regulating aspects more related to minerals policy-making.

The tax system provisions for mining and processing activities is complex. National, state, and municipal departments administer direct and indirect taxes: taxes on revenue, profits, and sales; taxes on products, services, and loans; and social taxes. The main taxes are the Imposto de Renda Pessoa Jurídica (IRPJ), a rent tax; the IPI, a tax on industrial products; the ICNS, applied to the circulation of products and services; the Imposto Sobre Operações Financeiras, charged on financial operations; the Financeiro Compensação Pela Exploração de Recursos Minerais, a financial-compensation tax for mining; and a number of social contributions, such as Finsocial, Fundo de Garantia por Tempo de Serviço, the family wage, and the Programa de Integração Social applied to sales. Tax reductions for IRPJ are given to firms operating in underdeveloped regions, and tax exemptions for IPI are given to exporters.

The current Constitution has advanced enormously in legislating mining operations and processing, as well as their environmental impacts. But the country seems to be caught in a transition between a very liberal law in the past and the present legal framework, characterized by instability and contradictions between different laws and policy instruments. In several areas, the constitutional decrees still lack regulations, mainly those needed to cover mining in native peoples' lands, the landowners' rights over the products extracted, and the relative decentralization of the administration of mineral resources. The new *Mining Code* has not yet been approved. Moreover, in practice, restrictions are only applied to foreign mining companies in frontier regions. Also, the past government explicitly intended to do away with most of the proposed restrictions on the operation of foreign capital.

Three main aspects of environmental laws have been somewhat neglected:

- A thorough cost–benefit analysis to help evaluate the feasibility of environmentally related investments vis-à-vis profit, by type of firm and by size;

- The need for uniform criteria across legal requirements, given the devolution of legislative responsibilities and administration to the states and municipalities; and

- An evaluation of the states' expertise to manage environmental policy and of the quality of their specific legislation.

As a result, entrepreneurs are faced with conflicting demands and are left much on their own to administer and harmonize conflicting interests and make them sector specific. This will be widely illustrated in the next section with case studies of specific firms in the bauxite, alumina, and aluminum sector.

Environmental behaviour of mining companies: strategies and responses

Five main phases can be distinguished in the history of the Brazilian bauxite, alumina, and aluminum sector.

- In the 1940s to 1950s, primary-metals production began, especially in the southeastern region (Minas Gerais and São Paulo). These operations were carried out both by transnationals (mainly Alcan) and domestic firms, such as Compañía Brasileira de Alumínio. Most inputs at that time were imported, and the local and smaller bauxite reserves, especially in the state of Minas Gerais, also began to be mined.

- In the 1960s, the big bauxite reserves in the Amazon were discovered, and the older firms continued to expand production and began to export.

- The 1970s saw greater expansion of aluminum projects oriented toward the domestic market while the largest projects in the North and the Amazon were being established.

- In the 1980s, however, expanded capacity at Alcoa's ALUMAR complex and at Compañía Vale do Rio Doce's (CVRD's) ALBRAS–ALUNORTE complex was directed mainly to exports. Meanwhile, the state-owned CVRD further activated its bauxite extraction and beneficiation project in Trombetas, developed by Mineração Rio do Norte (MRN).

- The 1990s, in contrast, seem to be oriented toward further rationalization of ongoing projects (especially those distant from the southeastern region), further verticalization, and reduced energy and labour costs. However, environmental measures only began to be applied in the 1980s, especially after the new Constitution.

The case of Compañía Vale do Rio Doce

CVRD entered the aluminum sector in 1974, but it wasn't until 1990 that this became its second main activity. In 1986, the firm supplied 72% of domestic bauxite production, and its primary-aluminum production (by ALBRAS and Valesul Alumínia S.A.) accounted for 25%. CVRD's aluminum-related exports constituted 57% of Brazil's total aluminum exports.

Four main periods can be distinguished in the firm's environmental policy. In 1956, the firm bought part of Mata Atlantica in southern Brazil for wood to build a railway line. But this unintentional initiative became, in time, a 21 700-ha private conservation area. In the 1970s, the company undertook its first projects with ecological aims. The projects related to feasibility studies for the Carajas project, the Minas Gerais mines, and the port of Tubarao; and to pollution controls for the Vitoria–Minas railway. In the 1980s, CVRD created GEANAM, an independent scientific-research area for environmental subjects and specializations. The firm formed seven internal commissions for environmental control, in different areas of the firm.

Two main types of environmental action were also defined: the corrective and the preventive. The first is included in all of the company's projects. The socioeconomic diagnosis of the Carajas project initiated the second. It was extended in the 1980s to environmental-engineering and natural-resource diagnoses, as well as scientific research, on specific topics, administered by university departments. The firm set up two laboratories for air and water control.

In 1986, the firm established the Superintendency for the Environment, subordinate to the Direction of Entrepreneurial Communication and Environment. The firm's subsidiaries set up their own environmental departments and internal environmental commissions (CIMAs) (17 in all). At present, the area of vegetation research is very developed within CVRD. Work is undertaken jointly by the Wood and Cellulose Superintendency, the Studies and Research Superintendency, and Rio Doce Florestas. Eight research centres function in the Amazon region, developing genetic studies of native and exotic species.

The 1990–95 environmental program established the following priorities: (1) the management of natural resources; (2) environmental engineering; and (3) the development of research and studies in the area. Fifty-three percent of the environmental investment was to be applied to carrying out the second priority, especially for pollution-control reforms in old operating systems and for preventive actions in new mining areas. In natural resources, 30% of the investment was to be used to develop a master plan and manage the wildlife-conservation areas of the firm, a total of 764×10^6 ha distributed in north, south, and central Brazil.

In 1980–88, the cumulative total investment in and operational costs of environmental management were 314 million USD, of which investments in infrastructure, equipment, and other improvements in the system accounted for 88.5%. Most of the firm's personnel for environmental tasks work for firms linked to CVRD and carry out reforestation.

Mineraçao Rio do Norte

MRN is the main producer of bauxite in Brazil. It is formed by domestic private, state, and foreign capital. However, in 1990, CVRD held 46% of the shares, with the participation of Alcan (24%); CBA (10%); and Billinton B.V., Billinton Metals S.A., Norsk Hydro, and Reynolds (5% each).

MRN's production is key to Brazil's status in the international bauxite market. In 1986, with its production capacity of 5.5×10^6 t/year, MRN was the fourth largest bauxite-mining company in the world; by 1988, it was ranked third. Moreover, together with Guinea and Australia, Brazil has some of the largest bauxite reserves in the world. In 1987–88, it contributed 8.6% of world production (UN 1988). The firm has, since its origin, directed most of its production to export. By the end of 1988, MRN had commercialized 38×10^6 t of bauxite, with 81% in the international market and only 19% in the local one. However, in 1990, it expanded its sales to the local market (28%), as a result of ALUMAR's having bought a significantly increased volume.

The firm's environmental department has two sections:

- The technical section, consisting of only one person, who is a senior employee, botanist, and special assessor reporting directly to the company's general superintendent; and

- The operational section, consisting by nine employees at the time of this research (November 1991): one general manager, one fieldwork contractor, one environmental engineer, three environmental technicians, two botanist auxiliaries, and one administrative auxiliary.

The department is part of a general department called Environment, Health and Security, and each section (including Security and Occupational Health, with eight people) has its own coordinator and is semiautonomous in decision-making. Actually, one of the problems expressed by our informants was that they tended to work too independently of each other.

The two sections in the environmental department proper have totally different functions. The operational section is responsible for revegetation, erosion control, pollution control, landscaping and gardening, greenhouse operations, meteorology, environmental inspection, and internal and external technical assistance. The technical section is in charge of formulating and following up on the environmental master plan (describing and relating the environmental risks of MRN's projects, problems, and solutions). This plan is revised every 2 years. The technical section develops or subcontracts relevant research and develops education and training plans for the environment in the firm and the community, carries out the internal monitoring, keeps a regular data bank, designs environmental rules and norms for the company, and establishes permanently protected areas. In 1989, the company persuaded the government to create a national forest of 429 600 ha around the bauxite reserves to protect the surrounding virgin forest. MRN maintains the physical integrity of this forest (for example, felling of trees and hunting are prohibited) and another 385 000 ha on its side that forms the Trombetas' Biological Reserve. MRN supports IBAMA's scientific research in those areas.

MRN has been concerned about the environment right from the beginning. The environmental department, run by the environmental assessor, was created in 1978. The firm has had a CIMA since about that time (as is the general policy in CVRD). The size of the department has varied throughout the years, but it was only subdivided into two sections recently. The plan for the future was to have the three subsections, including Health and Security, interact much more and bring environmental responsibility as much as possible to the shop floor. As part of the total-quality-control (TQC) plan, everyone becomes responsible for what and how they produce. The master plan can be translated into TQC measures for the environment. Beginning in 1992, the environmental assessor was to organize, according to the Environmental Executive Measures, a plan for environmental training of managers. Then, in turn, these superintendents and top managers were to train their people in each department, with support from the training department.

CIMA reports to the environmental department and the directorate and receives about 250 000 USD annually from the company. The main criticism of this type of organization is that it has a tendency to act as a political, instead of fiscal, body. Moreover, it acts as a normative body and not as an executive body, given that all firms within CVRD have their own environmental departments. The suggestion for the next few years is replace CIMA with another body that can really act as a fiscal body. It might include independent representatives from the community, such as householders, school teachers, and doctors.

Environmental costs can be divided into annual investment, which can vary substantially from year to year, depending on specific measures; and operational costs, which are more stable. Operational costs only vary according to the number of personnel in the department and the number of hectares deforested and reforested at the mine. The general superintendent's estimates of operational costs for 1992 are shown in Table 1.

Table 1. MRN's operational costs for environmental items, 1992.

Item	Operational cost [a] (USD)
Deforestation costs (including subcontractors for felling trees) • Average: 135 ha/year @ 1 300 USD/ha	175 500
Area preparation (including reforestation, leveling hills, returning topsoil, planting, replanting, breeding plants, combating ants) • Average: 80 ha/year @ 3 200 USD/ha	256 000

Source: Interviews.
Note: MRN, Mineração Rio do Norte; USD, United States dollars.
[a] Estimate.

Table 2. MRN's environmental investment, 1992.

Item	Investment [a] USD
IBAMA agreement to maintain the national forest and biological reserve	350 000
Reclaiming Batata lake	300 000
Erosion controls and landscaping	400 000
Small projects (such as for air-pollution control)	200 000

Source: Interviews.
Note: IBAMA, Instituto Brasileiro do Meio Ambiente e dos Recursos Naturais e Renováveis (Brazilian institute of the environment and renewable natural resources); MRN, Mineração Rio do Norte; USD, United States dollars.
[a] Estimate.

Total expenses for the year were about 1.6 million USD, which is in turn 0.41% of the original investment in the Trombetas project (390 million USD) and only 2% of the profit for 1991 (60 million USD). This was one-third of the environmental investment planned for 1992. Informants among top management supplied information on investment planned for 1992 (Table 2). However, sometimes there are extraordinary expenses related to reclaiming degraded areas. In 3 years, MRN had to spend 70 million USD, about 1 year's profit, to transport the washing plant to the mine site, partly recover Batata lake, and build new lakes (near the mine site) for the disposal of red mud from the extraction and washing of bauxite.

The master plan was formulated in March 1988, before the relocation of the washing plant (November 1989). So some of the environmentally degrading actions were already partly stopped. Although it was said that the plan is revised every two years, I did not have access to these revisions, if they were available.

Table 3 shows the main problems in the Trombetas' environment and their causes. The master plan points out that the worst problems are found in the area bordering Rio Trombetas, where one finds the industrial installations and urbanization. These degrading actions covered 1 752 ha in all by March 1988: 1 000 ha (57.1%) from Saracá mine (although 315 of 700 ha was reforested), together with the railway. The second important areas were those eroded by red clay from bauxite washing: 418.5 ha, or 23.9% of degraded areas.

MRN's reforestation projects seem to be the most advanced in Brazil (and can even be considered an example for the other mining companies, says Alcoa's environmental manager at Poços de Caldas). The environmental assessor made a detailed study of the species of the region and determined the format that reforestation must take, in what would otherwise be a rain forest. The format follows a

Table 3. MRN's classification of main environmental problems, 1988.

Problems	Environmentally degrading action
1. Deforestation, soil degradation and erosion, and river erosion	• Excavation in the plateau • Railway (30 km) and urban site building (100 ha) • Bauxite tailings sent to Batata lake after washing process
2. Water pollution	• Effluent from industrial area sent into Agua Fria and Fundao channels • Solid materials from degraded areas sent to the channels by rainfall • Effluent from offices and laboratories dumped into Trombetas river • Sanitary sewage • Oil accumulation in 5 km
3. Air pollution	• Dust in mine entrance • Dust on plateau • Dust and toxic gases in industrial area
4. Sound pollution	• Noise in industrial area • Unregulated transport and vehicles • Social events

Source: MRN (1988).
Note: MRN, Mineração Rio do Norte.

natural process of succession of species, considering especially those that will bring the fauna back, such as fruit trees that the local birds enjoy. The birds will propagate, via seeds in their excrement, new trees in the region. These innovative reforestation criteria are based on the way the forest grew originally — very different from the type of reforestation that focuses on fast-growing, nonlocal trees or aggressive plants and grasses and seems to aim at merely maintaining a green appearance. Ninety native species and about 12 exotic ones are used in replanting. After 10 years, when the soil has been rebuilt, shade-loving forest species will be introduced to finalize the reforestation process. The cost of this type of reforestation was estimated in 1989 at 2 500 USD/ha, which is 0.07% of the direct costs of the mine. In 1991, bauxite costs (shipping price) were about 10.5 USD/t, whereas the year before they were higher (13.64 USD/t). Almost half (500 ha) of the mined land was already reforested at the time of this research.

MRN bauxite deposits cover 40 000 ha of its 420 000-ha concession. At present, on average, 70 ha is mined annually. The wood from the rain forest, which used to be cut and burned, is nowadays used commercially. This has been enforced by an IBAMA law, and the company is still developing its plans for recutting the wood from trees felled near the mine and establishing the commercialization process. This process is not so easy because of the distance of the place from other populations and the high costs of transport.

The recovery of Batata lake, the site of one of the most denounced environmental problems in Brazil, needs to be discussed further. Since the beginning of the reclaiming of silted areas from this lake, different researchers from the University of Rio de Janeiro have evaluated the programs and recommended solutions. Professor F. de Assis Estevez did research on models to establish a new ecological equilibrium in the eroded regions. The main aim of such work is to recover the part of the lake affected by residue slurry, but a second objective is to help train scientists specializing in the recovery of tropical lakes. The erosion took place on the sides and in the western portion of the lake, and it had impacts on the flora, the fauna, especially the fish, and the hydrophilic vegetation. The extension of the eroded area is 318 ha of Batata lake (including the buried red mud); 92 ha of the small river or arm called Igarapé Caraña, which was being recovered; 5 ha of the Agua Fria branch; and 3.5 ha of Fundao Igapó. Solid effluent in these rivers ranges from 6 to 9%. The only impact on water quality is a higher turbidity where the water mixes up with the red mud; the clay itself is nontoxic. Analysis of water samples gathered at four stations (the first three affected by the red mud and the fourth without residue) show a smaller concentration of nutrients, such as phosphorus and nitrogen, at the first three points (MRN 1990).

An indirect problem developing from this change in the ecosystem is the exposure of the soil to the erosion produced by heavy rains in the region. This lack of protection for the soil, in turn, inhibits the permanent recycling needed by the nutrients in the biomass, which maintains soil fertility. In sum, the soil has lost fertility in the whole of Batata lake and surrounding area. However, the research projects focus on the creation of an organic substratum in the lake, with characteristics similar to the natural one; the revegetation of the whole area; and the recolonization by fauna (native species, especially fish). Two other problems pointed out during fieldwork on soil erosion have been the erosion of the sides of various rivers and the landslides and loss of forest from the slopes of the plateaus in the early stages of bauxite mining. The latter is being controlled by leaving a 5-m belt of bauxite at the margin of each plateau. The first of these two issues has not been satisfactorily tackled.

The ALBRAS–ALUNORTE complex

In 1974, as a result of an agreement between CVRD and the Light Metal Smelters Association of Japan, feasibility studies were developed to establish a project in the municipality of Barcarena, in the state of Pará, to produce alumina and aluminum. But because of delays in negotiations with the Japanese counterparts (financial, machine, and technology suppliers), its first exports of primary-metal ingots did not take place until 1986. That year, as a result of a lack of capital, the construction of ALUNORTE had to be temporarily stopped. Until 1990, ALBRAS had to rely partly on the imports of alumina and partly on supplies from ALUMAR.

ALBRAS — ALBRAS was conceived as a project oriented toward the export of primary metal; CVRD had an explicit policy of not competing for the internal market with the former local suppliers of aluminum. ALBRAS' environmental policy derived from CVRD's general policy for the Amazon-region ecosystem and was monitored by the local body, SESPA, the State Secretariat of Health of Pará. So far, the main problems faced by the firm have been the following:

- *Fluoride gas emissions from the open prebaked cells in potline 1 of the smelting area* — Thirty percent of the gases generated in this reduction line are still treated by the wet-scrubbing system, which generates mud residues with fluorides. These have to be deposited and treated, a difficult task because it can provoke liquid pollution, especially in a microregion so full of small rivers. To avoid the generation of liquid residue, the cells are being closed off and their feeding system is being changed.

- *The need to change all gas treatments to dry-scrubbing, also in the anode baking factory* — These changes are also under way.

- *Damage to the vegetation surrounding the factory related to excessive levels of fluoride deposits* — Measures are being taken to evaluate fluoride levels, but there is not a well-organized program. Every month, SESPA monitors these levels at eight points distributed between 1.0 and 8.5 km from the industrial plant. Similar problems in the fauna of local small rivers have been detected, but they have not been as thoroughly examined.

Interviews showed the problems had been detected but had not been thoroughly faced. Two reasons were given:

- The lack of personnel to tackle the problems, as a result of the complicated organization of the firm's environmental planning; and

- The lack of adequate prevision of effects because these issues had been loosely dealt with in the past.

The past is now coming to light. Flaws are informally attributed to inadequacies of the technical assessments of the past, on which the company had relied heavily; the problem was that the Japanese assessors had been unfamiliar with the climate and geographic conditions of Brazil. Besides these problems, observations from fieldwork showed that hazardous wastes and noise pollution, along with workers' health, would probably require closer attention. Quantitative information on noise pollution and health aspects is also scanty. This may be considered another indicator of the possible significance of the problem.

ALBRAS' organizational chart for the environment shows 150 employees. CIMA, established after the firm was set up, has representatives from the state and municipal governments, the Federation of Industries of Pará, and the Compañía de Desenvolvimento de Barcarena (Barcarena development company), responsible for the port's construction. CIMA dedicates a lot of its effort to environmental-education campaigns. It is more a political and diffusion body, although in the past, CIMA made some major decisions regarding policy implementation, such as the establishment of controls for emissions affecting water supplies near the plant; and decisions on the form of treatment and dumping of solid residues. It was also decided to set up six air-sampling stations for air-quality measurements. CIMA must regularly survey the results of the fume-collecting system, environmental hazards in transportation, industrial waste, sanitary sewage system, reforestation plans, and environmental-laboratory functioning. CIMA is not, however, an executive body.

Environmental investment was estimated at 120 million USD at the completion of phase 2 (August 1989), but the firm had not yet systematically calculated operational costs at the time of this research. These costs are for air- and water-quality monitoring. Air-quality control is done at the soil level and at the level of atmospheric emissions. The plant has a network formed by four automatic stations and 20 monitoring points. SESPA collaborates in this area to fix sulfation rates. The water-quality plan is rather limited and follows the specifications established by SESPA. But considering the dimension of the problem in a water-rich area (according to number of river branches and level of rainfall), it seems low. According to such programs, the standards reached in air emissions, liquid effluent, and vegetation were for 1990 (year for which information was available) in agreement with those fixed by the federal and state governments. However, the agency's local headquarters in Barcarena were not functioning.

ALUNORTE — The ALUNORTE project was initiated in 1980 and continued successfully until 1983, when it met adverse international market conditions. Then it stopped altogether in 1986, when NAAC froze its participation. However, a preferential agreement was reached in 1988, whereby ordinary capital and part of the already-granted financial aid became preferential shares. Then, with favourable market conditions, the project was reinitiated, but as a national private-capital firm. CVRD has its control and management and is looking for new potential partners. In August 1990, total final investment for the project, including what had already been paid, was estimated at 806.4 million USD.

During the interviews, environmentally oriented investment was estimated with the help of specialists in the field. Initial direct investment in environmental controls — including construction of tailings for red-mud slurry, urbanization and landscaping, sand and mud filters, laboratory equipment, and red-mud pumps — added up to 14 820 USD. This would be only 1.84% of global investment and 4.04% of total direct investment. ALBRAS is expected to provide much of the environmental experience and organization of ALUNORTE. However, the actual details of its functioning are not yet entirely clear. Information obtained from the interviews shows that beyond introducing adequate controls within production, little is really being done for the environment in this implementation phase, nor is the topic really seriously thought of or debated in management. Lack of advanced training on the issue is also among the flaws in this phase, especially given that much emphasis was placed on general training before project implementation.

Valesul

The history of Valesul begins with a proposal from the Reynolds' group to build an aluminum smelter near Rio de Janeiro. In a phase of high imports of primary metal and delays in the construction of the ALBRAS–ALUNORTE complex, the Brazilian government decided to take over the initiative. The government controls Valesul by having CVRD's majority of shares in the new firm and defining it as a public enterprise. The firm was created in 1976, but the participation of the Reynolds' group decreased over time. Subsequent operations only started in 1986. One of the main reasons for the interest in setting up this project was the possibility of locating the firm near the main urban centres with industries that use the primary metal, mainly Rio de Janeiro, Minas Gerais, and Sao Paulo.

The Environmental and Laboratory Division was established basically to execute the decisions made at a more global level by the firm. It was formed in 1991 with 7 permanent employees, 26 subcontracted workers, and 1 autonomous worker. The distribution of employment by type of qualification, form of contractual agreement, and task shows that 50% of the personnel carry out reforestation and 36.4% take care of residue treatment and conservation of green areas. For liquid-effluent treatment, there is only one operative. Finally, these staff constitute only 4.3% of total Valesul employment. Permanent, skilled employees (chemical engineers and technicians) constitute only 0.92% of total employment.

The environmental master plan was directed mainly to implementing a green area around the industrial complex in 1989. The general superintendency of CVRD for the environment, together with Rio Doce S.A. Forests, designed and established the implementation and maintenance costs of that endeavour. The latter company could draw on its previous experience at Tubarao and was hence responsible for technical support. But Valesul was to provide the operational human resources, materials, and equipment. The aim was to isolate the pollution sources, improve the scenery, and make the environment comfortable. CVRD's greenhouse at Linhares (state of Spiritu Santo) provided the species to be planted. By the end of 1989, 72 326 plants of 69 species had been planted in the following proportions: pioneers, 77.7%; secondary, 6.4%; fructiferous, 9.3%; and ornamental, 6.6%. Until then, the total costs of the initiative had been 38 060 USD, a cost per plant planted of 0.52 USD. In June 1991, this cost increased to 0.63 USD, including maintenance, and this made the initiative totally feasible. Other aluminum plants in Brazil have taken the green area as a model.

Other programs for the environment involve, first, monitoring the plant's pollution levels, done jointly with Fundação Estudual de Engenharia do Meio Ambiente (FEEMA, state foundation for environmental engineering), with parameters established by FEEMA in February 1985. Monthly and yearly emissions of total particulates, as well as total fluorides, are significantly lower by international and FEEMA's standards, with more efficient treatment of particulates than of fluorides. Second, there is a plan for the deposition of solid wastes, spent potliners, and dross. Although the problem has been studied sufficiently, with environmentally sound solutions either proposed or under study, not much has actually been implemented. In 1988, the Environmental Laboratory Division started coordinating efforts in this area, but lack of investment or political decisions have somehow delayed this program. Several important studies for the recovery of solid waste are under way at local universities and other firms, which will provide input to possible future solutions.

General information from the firm's archives showed that Valesul's cumulative investment in the environment before 1989 was about 28.42 million USD. This included the purchase of dust-collection equipment (by Brazilian standards, an advanced pollution-solving technology), the creation of a solid-waste facility, landscaping, and the purchase of necessary equipment for the environmental laboratory. Environmentally oriented investment has a constant tendency to increase.

For 1990, the operational costs and investment chart of the firm showed that investment in environmental protection for this year was minor, only 12.83%

of the total environmental expenditure (1.304 483 million USD). Almost the total capital is used by the firm for the reproduction of its operations, on a scale similar to that of the year before. Moreover, 75.74% of operational costs were in the smelting section, and 77.84% of investment was used by the laboratory to purchase equipment and accessories for pollution-control operations.

The study of CVRD's firms' environmental protection showed that much remains to be done, especially in water-quality control, dust and solid-waste deposits, and recycling. Workers' health and safety measures are also deficient.

The case of Alcoa

By the end of World War I, Alcoa had become a model of large-scale integration, from the mining of ore to the production of finished aluminum. With World War II, aluminum demand doubled, along with Alcoa's production capacity. Postwar demand for peace-time aluminum products helped the company grow still more in the 1950s and 1960s. It also started an aggressive program to sell aluminum on the international market. Today, Alcoa is the world's largest aluminum company, with activities all over the world. The United States has 105 of its operations. Brazil follows, with 15, and Mexico and Australia have 13 each. The firm is vertically integrated in all production stages and in raw materials, energy, and transport. But a clear-cut division exists between mainly local operations and those abroad. Third World operations involve the main bauxite mining and alumina and primary aluminum production, and these are on the rise.

Since the 1920s Alcoa has taken some type of action for the environment. Its basic concerns have included making environmental control compatible with economic growth. For this reason, it tries to collect and recycle all the materials it can within the production process and to incorporate controls into this process. Chemical materials, if recycled, reduce possible discharges of different types of liquid effluent. The technology has mostly been developed by Alcoa in the United States to satisfy both aims, especially since the 1970s, when controls tightened in the United States. The *Clean Air Act* amendment passed in 1970 to control air quality led to the creation of the Environmental Protection Agency, which established types and standards of pollution control. Alcoa increased its environmental investment from 5% of invested capital in 1971 to and 14% in 1979. This was a consequence of pressure from environmental movements of the late 1960s and legislation passed in the 1970s. In the 1970s, Alcoa spent 288 million USD on environmental protection alone. In 1981, Alcoa in the United States invested about 63 million USD in capital expenditures (9% of total capital investment) in the environment and 49.5 million USD in operating expenses for environmental needs. But the company recovered 43 million USD worth of materials to be used again.

Smelting is a good example: in 1981, the value of materials recovered through effective control technology in the smelting division exceeded 35 million USD, not enough to recover all of the operating and capital costs required to install, run, and maintain the equipment, but certainly a sizeable sum.

Alcoa in Brazil

Alcoa has operated in Brazil since the 1960s, and in 1970 it started its primary-metal production at Poços de Caldas in Minas Gerais. But the company Alcoa bought had been producing bauxite locally since 1930. In the late 1970s, Alcoa expanded its activities in the Poços de Caldas area, buying an electric-wires plant and installing an aluminum-powder plant. In 1980, it also began the country's most ambitious alumina and aluminum project, the ALUMAR complex in São Luiz, in the state of Maranhao. This was mainly to recover from setbacks to expansion already obvious at the Minas Gerais' operations. This process led to significant sales growth in the domestic market, opening of new export markets, and launching of new products in lamination, aluminum, metallic structures, and car bodies.

For general guidelines and execution, Alcoa in Brazil has differential managerial levels, complementary, but with different functions. The Corporate Department of Environmental Issues reports directly to the firm's local directorate. It has under its charge the formulation of policies and objectives for the environment and articulates with the departments in the plants. They in turn develop daily controls. They coordinate their work with that of the CIMAs, which are formed by representatives of all departments in the plant. In conjunction with the environmental department, they periodically analyze the relevant problems.

Table 4 shows the principal activities of the Corporate Department of Environmental Issues in São Paulo. The programs for the evaluation of environmental impacts, the audits, the plans for the prevention of environmental accidents, and the plans for environmental monitoring are the main executive vehicles of the plants' environmental departments and managers.

Alcoa (Poços de Caldas)

In 1967, Alcoa organized the firm Compañía Mineira de Alumínio (ALCO-MINAS), which had Alcoa as almost its only shareholder (99.9% of capital) and the support of the state government, through the Instituto Desenvolvimento Industrial (institute for industrial development) and the Banco de Desenvolvimento do Estado de Minas Gerais S.A. (development bank for the state of Minas Gerais). In 1980, Alcoa reduced its shares to 67.9%. But in 1982, it reinvested. By 1984, it recovered the position of control it had had at the beginning.

Table 4. The Corporate Department of Environmental Issues of Alcoa: principal tasks.

Type of activity	Tasks
Normative and political	• Establish and implement aims • Revise, interpret, and provide alternatives to environmental legislation • Represent Alcoa's position on the environment to private and governmental organizations • Assess its factories to obtain legal licences
Technical and diagnostic	• Assess factories in the implementation of environmental controls • Promote and follow up on monitoring programs for air, water, soil, and vegetation • Promote exchange of technical information with government bodies • Revise impacts of expansion projects or new investment to determine equipment needs for environmental control and monitoring • Research and determine the possible environmental impacts of different projects • Assess the firm's solutions to environmental problems

Source: Research.

In 1970, ALCOMINAS was the first local producer of aluminum, and it also produced alumina. This was Brazil's first project to include investment for the environment.

The environmental policy and operations for alumina and aluminum production follow the same pattern as those of the ALUMAR complex, described later, so special emphasis will be placed here on those developed during bauxite mining. However, relevant control and measurement standards were not obtained during the field research on activities in Poços de Caldas. Probably the older technology in use (Soderberg type) made this material unavailable.

The firm is on a plateau near the city, which in turn is 180 km from São Paulo. The plateau is circular; it is 1 200 – 1 600 m high and about 400 m high at the periphery. Bauxite is concentrated in the northern part of the plateau. Alcoa works yearly on 5–10 ha and has an overall reserve of about 356 ha. Topsoil from the area to be mined is stockpiled until it is needed for recovery and reforestation.

When the mining is completed, terraces are created in the sloping soil. The terraces allow rainfall to concentrate but stop the soil erosion that would otherwise occur. Topsoil is then distributed to the mined areas to prepare for planting. Calcium and chemical fertilizers are added because high acidity levels and lack of nutrients have been discovered. Drainage systems orient rainwater to reservoirs built in the mined area. This increases the soil's capacity for absorption and avoids erosion.

At present, the reforestation program uses native species and has the input of professional staff from the University of Campinas in São Paulo. This is the modern approach used by Alcoa; in the past, reforestation depended on commercial and homogeneous species, such as eucalyptus. Reforestation costs depend on the specific characteristics of each area, but since 1979 they have been about 0.00065 USD/t of ore mined. This is quite a small relative cost, even if, considered in absolute terms, the recovery of a whole mined area seems costly. For example, in 1982–83 an area of 4.42 ha (a year's mining work) was recovered for 7 425 USD. But as it produced 225 234 t, the relative cost was of 0.00145 USD/t. The firm has been developing these operations since 1977 and has already recovered 135 ha. Labour costs in operational costs for each stage of recovery (terraplain, drainage, revegetation, maintenance, and greenhouse) were 55.80% of total costs. In sum, the cost of recovery per tonne of bauxite is about 1% of the cost of mining it and transport it to the alumina factory; that is to say, it is economically feasible.

The ALUMAR complex

The ALUMAR alumina and aluminum complex is on São Luiz island (which has an area of 900 km^2), in the state of Maranhao. The complex occupies 1 700 ha. It is part of the Northeast Region in Brazil, but legislation made it part of Legal Amazon in 1967 (decree 200). This means that it has a right to all the subsidies for economic projects in that region, but it is also subject to all the controls. The area has a very rich hydrological system; the most important rivers for ALUMAR is Rio dos Cachorros and its tributaries, Igarape Andiroba, Pedrinhas, and the little Coqueiros river, which is the main drainage system for the plant. Only 34.5% of the 1 938 km of rivers is navigable, because of the lack of more adequate infrastructure. Mango trees occupy 15% of the island. However, many of the previously existing species were devastated by the activities of people, and this reduces the variety of animals for human consumption and accelerates the area's erosion.

At ALUMAR, the concept of incorporating environmental controls into the production process implies a major initial and gradual investment. The environmental department's budget for 1990 was about 2 million USD, used mainly for operating costs, such as training, wages, landscaping activities, and laboratory controls. During the fieldwork for this research, investment in 1990–92 for environmental activities at ALUMAR was estimated at 55.14 million USD, distributed as shown in Table 5. Levels are high because they include the rehabilitation of one of the lakes for deposits of red mud.

Table 5. ALUMAR's investment in environmental control and recovery, 1990–92.

Items	Investment (million USD)
Building	52.8
Stocking building for SPL	
Bauxite residue lake 2	
New ash deposit	
Expanding or creating systems	1.3
Alumina-transport system	
Expansion of sewage system	
Gas-treatment system in anode baking	
Recovery of bauxite-residue lake 1	
Irrigation system	
Opacity-measurement system	
Technological development	0.1
Recycling of SPL	
Environmental recovery	0.9
Recovery of west area 261 (10 ha)	
Landscaping plan	
Greenhouse	
Degraded area recovery	
Environmental park	
Emergency	0.1
Emergency-action plan	

Source: Data from ALUMAR's archives.
Note: SPL, spent potliners; USD, United States dollars.

In 1984, the consortium built and started operating a lake for bauxite residue, covering 15 ha. It had been designed to be used for 4.5 years, but it was enlarged to increase its capacity. The artificial lake is formed by dykes built with local solids, and it has a double impermeable membrane of polyvinyl chloride, 0.8 mm thick, that covers the base of the lake and the sides of the dykes. This membrane avoids the infiltration of residue into the superficial waters, which could thus be contaminated. In 1990, a new lake was put into operation. Investment in this was about 27 million USD. The lake has a capacity for 4×10^6 m^3 of residue and took more than 1 million working hours to build. As this new lake begins to receive residues, the older lake no longer collects effluent, but it is still not deactivated. In 3 or 4 years, through rainfalls, it is expected to have a natural cleaning process. Only when its water is considered environmentally sound will it begin to be recovered. This is an ongoing process.

The environmental effects of the construction and operation of this complex were studied for 3 years before the project was implemented. This study had the participation of the environmental department of ALUMAR and Alcoa, the Hydrobiological Laboratory of São Paulo University, and the State University of Maranhão through contracts and consulting agreements with ALUMAR. It is very interesting that the evaluation of environmental impacts anticipated the Brazilian dispositions that began to be applied in 1986; ALUMAR started operating in 1984.

Total fluoride emissions and particulates were above international standards in the potroom in 1990. For fluorides, this continued into 1991, although at a much lower level. The environmental department is working hard to limit them, especially those related to the operators' inadequate handling of the equipment, such as when they keep pots open for too long while they perform certain functions. Because of this, a third instrument, a spill-prevention plan, was established to identify and inventory all potential spills. Accidental spills of oil or hazardous products resulting from operating failures or breakage of tanks or ditches can cause critical pollution problems for the environment and the workers. As part of this plan, every morning and before each new shift, 10 min is devoted at the ALUMAR complex to discussions of hazardous working-environment issues.

Both cases

In both the case studies (CVRD and Alcoa), the following environmental-control procedures were used:

- An EIS, whose monitoring measures are mainly applied to fluoride emissions and particulates;

- Partial land and tailings recovery;

- CIMAs; and

- Some form of environmental training (although here there were substantial differences between the content and extent of the programs applied by each company).

What also varied between companies was their knowledge of the best practice for certain environmental controls and their applications. The multinational firms had wider knowledge. The companies also used basic health, safety, and accident-prevention monitoring, which will be further discussed in the next section. However, very few companies had sufficient water-quality controls, locally specific land recovery, or full reforestation. Dust controls were virtually nonexistent, and so were noise- and heat-level controls. The companies did very little recycling of by-products. Solid wastes from alumina and aluminum production were not thoroughly protected in stocking. Finally, monitoring of local flora and fauna was inadequate (especially near tailings). Moreover, environmental-laboratory technology tended to be either outdated or insufficient.

Innovation and human-resource development

This section explores the main approaches to innovation and technical change, employment, training, and health and safety in this sector. Whenever possible, I will be drawing on more detailed information from the case studies.

Technology

Technological change in the domestic industry is represented by two main stages in the smelting process: Soderberg smelting and prebaked smelting, which is a more environmentally sound technology. The two types of smelting differ in the type of anodes employed: prebaked smelting involves a multianode; the Soderberg continuous anode, which is baked *in situ*, uses the heat generated by an electrical current passing through the cell (UNEP 1986a). However, a series of incremental innovations is a later part of either stage, although this is more characteristic of the prebake technology. Both types of pots can be closed or opened, hooded or unhooded, which makes a big difference in the gas emissions in the workplace and in the atmosphere. The types of cells also differ in the ways the anodes and alumina are fed into the process and the ways the anodes are replaced. The less the inputs are in contact with the atmosphere or the operator, the cleaner and safer is the technology. Other incremental technical changes have taken place in the size of the smelting pots, process controls, magnetic stability in the pots, and operational and control parameters in alumina feeding.

Since 1930 researchers at Alcoa in the United States have searched for an effective system to collect and recycle fluorides. In 1967, the first unit of Alcoa 398 technology was put into operation. It is a dry system of fluoride-fume treatment and recovery; 99% of fume-fluoride content can be collected and recycled in the smelting process. New plants were built with that process, and almost all older plants have had their systems replaced. In 1979, people thought that if the system was changed in all plants, this would produce savings of 80 million gal (1 gal = 4.546 L) of water and reduce fluoride supplies by half. In 1975, the 446 process was designed, a dry-scrubber to remove tarry materials with fluoride from the carbon blocks in the baking process. Other less extended developments have included, in 1979, the new smelting technology to produce aluminum without cryolite, avoiding fluoride emissions altogether. With that technology, dry-scrubbing is still needed for other contents of gases (particulates, chlorines, and other compounds). The policy with respect to water is conservation plus reuse. Both the 389 and 446 technologies help to reduce water pollution insofar as they do away with wet-scrubbing. By replacing 75 wet-scrubbers with dry-scrubbers, Alcoa reduced water use in 1967–90 by more than 5×10^6 gal/d. Another advance has been an improved treatment system to recover almost 90% of waste water for reuse; this is achieved through the elimination of dilute oil. Other Brazilian firms now use similar prebake technology and dry-scrubbers.

All aluminum smelters in Brazil have foreign technology, although some incremental technical changes have been designed and applied locally. However, larger projects make no provision for developing local technology, except of course in terms of some of the architectural designs that are so central to the industry. Research and development (R&D) institutes and centres are scarce in this industry; any that exist are found mainly locally in ALBRAS, Alcan, and CBA. The main areas of R&D have been in process automation, optimization of technical parameters in smelting, selection of more efficient materials, reduction of energy consumption, etc. On selected topics, the industry has professional input from R&D at local universities.

Innovation is largely embedded in imported machinery. However, plenty of scope exists for incremental technical change, especially in the smelting rooms. These can be environmentally oriented, as this also reduces waste of materials, thus making it economically feasible. Still, power failures, which are not infrequent, can reduce the productivity and environmental soundness of increasingly cleaner technologies. This seems to be an urgent problem. Lack of communication between the technical and production departments and the environmental and health and safety sections also hinders the firms' implementation of a more integrated innovation policy; it also reduces environmental quality. The role that training can play in the developmwnt of a more integrated effort will be discussed in the next two sections.

Employment

Table 6 shows the distribution of direct employment in the aluminum industry by type of activity and type of job for 1990. Total employment was 66 780 people, with the majority (84.4%) occupying technical positions. The latter include blue-collar workers, as well as technicians and professionals. The table shows the importance of the independent manufacturers as employers: 56.3% of employment was in that subsector, whereas 40.3% was in the primary-aluminum integrated firms, those mainly studied in this research.

Table 6. Brazil's direct employment in the aluminum industry, 1990.

Subsector	Administrative (n)	Technical (n)	Total (n)	Total ()
Primary-metal integrated	4 879	22 032	26 911	40.3
Secondary	380	1 873	2 253	3.4
Independent manufacturers	5 169	32 447	37 616	56.3

Source: ABAL (1990).

The evolution of employment in 1984–90 by type of job shows an increasing share of technical jobs in general employment. Technical jobs were responsible for all employment growth in 1984–90. However, in the 1990s, the number of jobs has decreased in absolute terms. This reflects the policy adopted by the large corporations in this sector. Among integrated primary-metal producers, the policy was mainly to decrease administrative jobs so as to reduce production costs and maintain competitiveness in the international market as metal prices fell. This was partly a result of specific management techniques, such as TQC. In contrast, technical jobs remained quite constant in this period in the category of primary-metal integrated firms and rose more regularly in the other two subsectors. Brazil is one of a group of countries that have low wage costs; the others are Australia, India, South Africa, and Venezuela. Social wages form a large part of average wages, although the proportion is not necessarily higher than in other producing countries. Furthermore, during the recession, the producers have shown a preference for hiring workers on a subcontractual basis. Social costs are then to be borne by the subcontractor, rather than the employer. However, subcontractors are subjected to fewer government controls.

Other information on employment and wages, published by the Instituto Brasileiro de Geografia e Estadistica (IBGE, Brazilian institute of geography and statistics), shows that in terms of number of jobs and wage levels the recession in the domestic industry has been stronger in the mineral-extracting sector. This has resulted in job losses and wage decreases (IBGE n.d.). Unfortunately, the criteria for collecting employment data in bauxite production in Brazil differ from those used in the aluminum industry. The DNPM data are the best available and are published in the *Brazilian Mineral Yearbook* (DNPM 1990). This source is built with periodic information filled out by the firms themselves. Employment was at 1 543 in 1989, with the vast majority being blue-collar workers, and this level remained stable for several years.

Strategies for training, health, and safety

All firms studied had their own training departments. Some of these training departments (such as those of ALBRAS and ALUNORTE) were even operating during the development phase of the project, before installation. Many of the engineers, technicians, and managers had been sent abroad to train or retrain at the mother company or with technology suppliers. Local retraining was usually done systematically, during operation. For example, by August 1989, already 92% of ALBRAS' workers were local people. They had all been trained in the plant, because no one in the area had had previous experience with aluminum or heavy industry. Delays in project implementation gave the compnay time to develop, well in advance, training plans to be used locally and in the national universities. However, technical personnel had to be recruited from outside the region and even from outside the state, sometimes from other CVRD firms. To initiate the project, the firm also had 60 Japanese technicians and aluminum specialists from Mitsui. However, it was harder to estimate what proportion of this training was related to the environment. In the environmental departments, usually only the heads had such training abroad; middle managers and technicians participated in national conferences and courses, but not systematically.

Alcoa had one of the most advanced training programs in the sector, and this can be illustrated using the employment and training practices at ALUMAR. In June 1991, permanent and subcontracted employment was at 3 358 persons, almost one-third of Alcoa's general employment in Brazil. Permanent employment was at 2 872: 68% were paid hourly; 15% were paid monthly; and 17% were hierarchical technical and managerial staff (15 in all were employed in the environmental department). Personnel had higher wages than similar workers in other regional industries, and their jobs were more stable than similar ones at MRN. Schooling levels were also higher than at other local firms. Moreover, ALUMAR did systematic and permanent in-plant training, especially in the most technical areas. It also regularly sent top- and middle-ranking managerial and technical staff overseas for training. Table 7 shows the firm's training budget, 579 374 USD, for 1991: of this, 8 011 USD, or 1.38%, was destined for personnel in the environmental department.

The company also had daily 5-min discussions of health and safety — a legal requirement. However, part of this time was sometimes used to discuss environmental issues. Training is one of the main methods for dealing with these issues — training in emergency-action plans at the level of day-to-day details and routines. At ALUMAR, a follow-up was also made: information on the number

Table 7. ALUMAR's budget for personnel training, 1991.

Area	Training costs (USD)
Operational management	96
Materials	20 003
Port	6 116
Human resources	45 751
Environment	8 011
Public relations	583
Accounts office	11 004
Quality	2 369
Security	26 379
Systems	19 091
Refining	39 823
Smelting	31 118
Ingot factory	2 638
Engineering and maintenance	25 072
Administration	1 012
Alcoa and subsidiary	923
Start-up	5 567

Source: ALUMAR's archives, Administraçao Industrial (1991).

of accidents and the attitudes of those involved was added to a data bank. For example, at ALUMAR in 1990–91 (until July 1991), of 18 accidents occurring in the calcination section, 5 (27.8%) were caused by human error. Of 116 accidents in the clarification section, 40 (34.5%) were caused by human error; 15 (12.9%), by spills in tabulations; and 14 (11.9%), by faulty equipment. The environment departments investigated each accident to classify it by degree of environmental damage, identify its cause, and establish corrective or preventive action. If human error ranked high as a cause of accidents in a section, a questionnaire was distributed to that section's operators. This was supposed to be a consciousness-raising experience and was considered useful in evaluating and eventually increasing — through training — the workers' knowledge of environmental and personal risks in their section.

Other departments at ALUMAR carry out other training activities complementary to those strictly oriented to the environment. I must at least mention general training activities and industrial-hygiene programs. The first considers, in a limited way, environmental training in the recruitment phase. All recruits are shown a film on environmental controls at the plant; for certain jobs, environmental aspects are included in initial training programs. Workers in some situations, such as potline workers, receive environmental training when certain pollution standards are not being met. The plant also promotes some environmental training with community groups, offering 1- or 2-d programs with open discussions of the topic, etc., and also gives technical assistance for setting up communal greenhouses. In the past, the company made some fairly high investments in retraining its process engineers and environmental engineers, either domestically or abroad. However, the recession drastically reduced its training investment.

Industrial-hygiene programs are crucial in this industry because one of its main problems is the need to improve the direct working environment, principally in the potrooms, at the anode-fabrication plants, and in some sections of refining. Heat and noise levels tend to be high at most stages of the production process. At ALUMAR, this is especially so, considering the location of the complex is already extremely hot and damp. Mean exposure levels by job class, as agreed internationally, are periodically compared with real levels attained in various sections of the plant. Two of the main difficulties are still the reduction of noise levels and human exposure to corrosive chemicals. For this reason, the plant provides a complete list of the main agents and the numbers of people potentially affected by their activities per stage in the production process. This is used to monitor workers and technology.

MRN has also developed a very comprehensive program to tackle health and safety problems, and the firm has been most interested in reducing work-related accidents, which are common in all mining activities. In 1991, a new program was developed to reduce the accident rate relative to 1990. The firm elaborated some policy instruments for supervisors to put into practice in various areas. The firm uses posters, talks, etc., to publicize the general security policy. Also of use are monthly meetings on health and security; daily 5-min security talks; auditing of insecure actions; preventive analysis for various jobs; and programed security inspections.

Tables 8 and 9 show the number of accidents among permanent employees and subcontracted ones at MRN between January and September 1991, along with those foreseen by the company for that period. The tables indirectly measure the level of risk of each accident according to two categories, that is, whether or not they resulted in lost time. Most accidents were without lost time (more than three-quarters of cases among both types of personnel). The company's predictions were somewhat higher than the results.

Table 8. Work-related accidents among permanent employees at MRN, January–September 1991.

Accidents	With loss of time	Without loss of time	Total lost time
Actual			
n	5	25	30
%	16.5	83.3	100.0
Predicted			
n	11	35	46
%	23.9	76.1	100.0

Source: Company archives.
Note: MRN, Mineração Rio do Norte.

Table 9. Work-related accidents among subcontracted employees at MRN, January–September 1991.

Accidents	With loss of time	Without loss of time	Total lost time
Actual			
n	15	121	136
%	11.0	89.0	100.0
Predicted			
n	32	142	174
%	18.4	81.6	100.0

Source: Company archives.
Note: MRN, Mineração Rio do Norte.

The number of accidents among the subcontracted workers was 4.5 times higher than among the permanent personnel. But the number of subcontracted workers (2 869) was more than double that of permanent workers (1 282) for that year. The data show that MRN's industrial security policy has much greater success reaching the permanent personnel. The mine is obviously the most risky place to work. The mine, the port, and the beneficiation area together account for 84.61% of all accidents. Fieldwork observations showed that the workers most affected by accidents are, in order of severity, maintenance personnel, blue-collar workers, and drivers. Accidents usually happen because mechanical parts bump inattentive workers or tools fall from their hands. Company estimates show that fingers are the body parts most frequently affected. This category accounts for 21.88% of cases; arms and face, 9.38%; and back and forearms, 6.25%. Other body parts are rarely injured.

In ALBRAS–ALUNORTE and Valesul, training in general and training on the environment and safety were less advanced than at the other companies studied. But conflicts arise in all companies over the articulation of training endeavours — between the general training offered by the personnel and training department and the more specific forms of training needed by the environmental and health and safety sections. Moreover, the environmental input does not inform all these training policies, which tend to function independently in most firms. Above all, hardly any systematic integration exists between the technology and production sections and the different training initiatives. This partly shows how little environmental training has been planned in this transition phase. Environmental training is done without taking a holistic approach to all sections of the company. The next section, however, analyzes indications that this might begin to happen in companies or in the government's future plans.

Company strategies and government policies for the future

Main conclusions and policy recommendations

The case studies illustrated, to the extent possible, the everyday environmental behaviour of two central Brazilian companies. The data from these two companies cannot be extrapolated to the whole sector, but the studies offer a sound illustration. From this, some trends can be derived regarding the relationship between environmental regulations and company behaviour (environmental management and technical solutions) and — although in a more limited way — between social pressure and company behaviour.

The case studies revealed the main forms of relationship between environmental regulations and company behaviour:

- Regulation, either local or home country in the case of transnationals, is a necessary but not sufficient condition for implementation of a sound environmental policy at the company level in Brazil.

- The level of environmentally sound behaviour varies between aspects of environmental planning, showing a set pattern across companies. In other words, some regulations are more systematically applied than others. This in turn is positively correlated with the direct economic gains companies obtain from obeying regulations and only secondarily with the existence or enforcement of the regulations or social pressure. For example, fluoride-emission control is heavily monitored across companies, whereas sound reforestation and riverbank erosion are controlled less uniformly.

- Regulations and social pressure are the main motivators for the companies to implement adequate measures against negative environmental impacts that are more visible to the wider public or that affect a broader population. Regulations and social pressure are interrelated: usually, general social pressure results in new regulations, whereas particular social pressure results in a more thorough implementation of rules, along with careful monitoring.

- Even in cases where environmental soundness in a company's behaviour is triggered by regulations established as a result of social awareness in the home country (as with Alcoa), companies can demonstrate time lags, resistance, or denial in carrying out the necessary adaptations to the local environment. This shows that the local environmental efficiency of some transnationals does not derive mainly from acceptance and enforcement of local regulations, but from the companies' greater technological and financial capacities, which are only in a limited sense a function of environmental regulations in the home country.

- Corrective maintenance predominates in this sector. This is partly a result of the characteristics of local regulations for the sector (recent, flawed, or contradictory, as explained earlier) and partly a function of the type of growth the government fosters.

- Even if regulation ix the more direct cause of positive changes in companies' environmental practices, flaws in the implementation of the laws and policy measures diminish the potentially positive impacts of environmental regulation.

- Not all transnational companies are at the forefront in the use of environmental-control technology or management in Brazil — in certain respects, the national laws do not lag the industry's environmental practices. For example, CVRD's behaviour, as shown by this study, is more advanced than Alcan's local practices, and the national laws on reforestation and land rehabilitation using local species are very convincing and adequate on paper, if not in practice. The main problems are establishing an adequate institutional setting and the mechanisms to monitor implementation and to make regulations effective; and creating political power to enforce the regulations vis-à-vis different economic groups. These problems are clearly illustrated by the scanty application of environmental regulations that directly affect the workplace. These regulations are precise and available, but they are not effectively applied. The workers have little power, and government practices do not support the workers.

Ultimately, then, environmental soundness depends not only on the instrumentation of environmental regulations and sound technical choices but also on having an institutional setting that favours their application and having the educational skills and political power to create these instruments. Without these, neither regulations nor technical and managerial solutions are enough, although they constitute an essential precondition.

Let us look now at the case-study results in more detail. The case studies showed that most of the companies had production-expansion programs under consideration. MRN intended to expand its production capacity of 3.5×10^6 t/year to 5.5×10^6 t/year and to supply the ALUNORTE plant once it started operating. ALUMAR of Alcoa planned, for 1993, to increase production to 950 000 t of alumina and 380 000 t of aluminum, almost doubling its 1990 capacity, both for export and for the supply of its domestic manufacturing plants. It was continuing its verticalization process. ALUNORTE was to begin its operations as the main supplier to ALBRAS. With this, ALBRAS was to expand its production of primary aluminum substantially, to 340 000 t/year. However, ALUNORTE intended to produce 1.1×10^6 t of alumina/year for the global market by 1996.

For most of these initiatives, the companies upgraded their technology, usually through incremental technical changes. These resulted indirectly in a cleaner production processes. The best international practices were adopted for new endeavours, such as that of ALUNORTE or the expansion of potline sections within smelting, at ALBRAS. These tended to be environmentally sounder. ALUNORTE even chose the Giulini system for disposal of red-mud slurry — one of the most advanced systems in the world. However, in the plants operating with older technologies (the Soderberg vintages), less was being done in both these directions, that is, toward the expansion of operations and toward technological upgrading with environmentally sound hardware. Partly, this reflects the limitations imposed by the technology; partly, the firms' lack of motivation to innovate. Moreover, local R&D for these technologies is scarce, and so is the knowledge on best practices.

Differential environmental experience was shown within the firms studied. In most, the controls, whether regulated by recent laws or not, were applied more in a corrective than preventive manner. In other words, they were considered and used after the technology had been selected and only when some of the problems presented themselves. This trend had been stronger in the past among firms with less experience in producing bauxite, alumina, or aluminum, such as some of CVRD's plants. Some multinationals, such as Alcoa, had had environmental regulations in their home countries earlier. These companies also tended to have a greater capacity to embed the necessary controls in the technologies they developed. Familiarity with production and R&D capacity made it feasible for some companies to plan ahead and develop a preventive policy from the implementation phase of a new project. These activities also made it somewhat easier for these companies to keep up with domestic regulatory standards. However, the companies tended to lack experience in the specific environment they were dealing with.

The general environmental laws applied to the sector are contradictory regarding standards and types of controls or they are highly specific to each state or region. This makes it complex to set up common guidelines across activities within one company or across companies. As the regulations are generic for mining or for industries as a whole and deal with air pollutants, chemicals, or water pollutants by type, entrepreneurs and environmental departments are generally left to decide their own activities. Fortunately, the Associação Brasileira da Indústria do Alumínio (ABAL, the Brazilian association of aluminium producers) has been able to draw on its past experience in the sector, systematize the legal information (ABAL 1991), and informally draw up some common guidelines. The association holds periodic meetings for environmental departments, a general conference, and annual training workshops to discuss environmental information relevant to the sector. This seems the right forum for generating a more preventive environmental approach for the whole sector. Alcoa's initiative is very strong in ABAL, and it exerts its influence widely, given its longer trajectory in the area.

However, ABAL has accomplished little toward devising policies covering solid waste and dross recycling in the industry; aluminum can and scrap recycling; or heat, noise, and reverberation reduction. These are the most prominent environmental hazards still largely unresolved in the Brazilian alumina and aluminum sector.

There is no body that is equal to ABAL or develops similar tasks for bauxite mining, exploration, and development. IBAMA tends to fulfil a more administrative role as a government environmental agency. Some research on the local flora and fauna, as well as on specific rehabilitation of degraded land, is being done at the Goelde Museum in Pará; at local universities all over the country; at R&D institutes belonging to the companies, especially those of Alcan, Alcoa, and CVRD; and in the local environmental agencies. The latter often hire the heads of the companies' environmental departments for their expertise. However, there has been little systematic integration of research results.

The case studies also illustrate how much the advances in formulation of the laws are not reflected in coherent implementation practices within the companies. For example, some government measures are applied bureaucratically, that is, without assessing the possible results of differing implementation schemes in particular environments. Conflicts between departments in the firms make it difficult to devise a more integrated approach to training staff to implement environmental measures.

The skills of personnel in the environmental departments also need to be upgraded. This is especially the case with national firms, which either have too few people dealing with the environment or too few professionals and many more subcontracted labourers in manual operations. On the other hand, manual and low-tier supervisors and technicians have no access to vocational-training programs with an environmental approach. Until recently, SENAI (the industrial vocational institution) has not considered setting up any new programs with this concern or including this concern to any substantial extent in its regular curriculum.

A lot more needs to be done in all sectors to protect the workplace environment. This also applies to current domestic environmental regulations. The case studies show how relatively little has been done to diminish the adverse effects of heat, sound, or reverberation in the workplace or, in some cases, the nearby neighbourhood.

The proposed (preproject) amendments to the environmental legislation in the Constitution will do little to increase the environmental soundness of the sector. These are the clauses commented on in the second section of this paper: they call for further administrative and legal centralization of controls, monitoring at IBAMA, and restrictions on participation of foreign capital.

The regulation of the use of native lands for mining endeavours also awaits further resolution. For this, the wider participation of interest groups is not only desirable but advisable. The public-audiences policy instrument for the discussion of RIMAs before project implementation has not yet been widely promoted by government regulating agencies. Despite this, it may prove an interesting tool for public awareness and participation.

The following main policy recommendations arise from this research:

- The need to emphasize a preventive versus a corrective approach to the environment among firms and in government policies and legislation;

- The need to develop a more systematic social policy at both levels, which may in turn benefit the environment and the different actors; and

- The need to further engage the various social agents in discussion and participation (for the sake of open dissemination of information) and in decision-making on environmental issues, as these tend to have a wide-spread effect at a local, regional, or national level.

A word of comparison

Using other studies undertaken for the Mining and Environment Research Network (made available from China and Ghana), I will briefly discuss some similarities and differences in the trends found in environmental management and land recla-mation, most especially as they relate to the aluminum industry and bauxite mining (see Acquah 1993; Acquaisie 1993; Lin et al. 1993).

First, in all three cases — Brazil, China, and Ghana — strong regulatory institutional settings were in operation when the case studies were developed. These settings differed in their time frame, the type of pressure that set them in motion, and the degree of linkages with local industry to make their policies effective. China had the oldest initiative, dating from 1973, followed by Brazil. The more recent initiative is articulated in Ghana's National Environmental Action Plan for 1991–2000. Although in Brazil and China, environmental initiatives at the national level derive from general social awareness and pressure, in Ghana they stem more clearly from the World Bank's guidelines as preconditions for loans. A typical example is the Ghana Environmental Resource Management Project, a 5-year plan to develop the necessary institutional and technical capabilities to combat soil and resource-base degradation. Industry linkages seemed to be far reaching in the case of China; not especially notable in the case of Brazil; and apparently just beginning in the case of Ghana.

Second, the cases in Brazil and Ghana were similar in having a double channel for environmental soundness: on the one hand, government monitoring and control; and on the other, corporate practices (that is, transferring to the Third World the environmentally sounder technology developed to meet First World standards). In Ghana, transnationals seemed to be doing this, but practices were not so clear cut in Brazil. Brazil was also more complex, as some industrial initiatives were outside the hands of the transnationals. Because the transnationals were operating under technology-transfer agreements, they had a lot of room for in-plant adaptations and incremental technical change, which can alter environ-mental soundness significantly. In China, initiatives were developed locally, especially in land reclamation for bauxite production, and this always involved a high level of local R&D and research *in situ* at each mine.

Third, China seemed to have the highest levels of R&D expenditure on the environment at national and sector-specific levels. China was also the only country that had established quota-based, or market-type, instruments for environmental control: since the 1980s treatment-control funds have come from the polluting organizations, rather than from the state. As we have seen, Brazil has an orientation in this direction, but not Ghana, where the command-and-control approach predominates. Also, in China, by law, 7% of technical-innovation funds must go to the struggle against pollution (Rothwell 1992; Skea 1993).

Fourth, none of these countries had policies disaggregated to a sector-specific level, leaving associations of producers to determine largely their own parameters and operationalization, with the exception of the land-reclamation programs in China. These were highly sophisticated and mine specific, usually with the objectives of making mined land arable again and achieving the standards of the local farmers.

Fifth, in all cases, the general focus of policy, in theory, was on prevention, but in practice the focus was on environmental controls for problems that had already surfaced — that is, on correction rather than prevention. In all cases, the main obstacle was the lack of managerial skills for environmental preservation. The literature in the three cases also indicated that best environmental practices could be attained with community involvement in policy-making. These are specific concerns and setbacks in attaining environmental soundness in developing countries, and they should be taken much more into consideration by those who are developing research and policies for sustainable development.

CHAPTER 8[1]

COMPETITIVENESS, ENVIRONMENTAL PERFORMANCE, AND TECHNICAL CHANGE: A CASE STUDY OF THE BOLIVIAN MINING INDUSTRY

Ismael Fernando Loayza

This chapter analyzes the links between competitiveness, environmental perform-ance, and technical change in the Bolivian mining industry. It develops a dynamic economic model of the mining firm and tests it empirically using a multiple-case study of four Bolivian mining companies and seven mining operations. This model combines an economic theory of depletion with a theory of pollution and com-prises two sets of equations describing investment behaviour and pollution per unit of output. The model analyzes the ways companies compete through technical change and illustrates how competitive companies increase their production capac-ity and technological capability over time.

The principal finding of this study is that a mining firm's dynamic ef-ficiency significantly affects its internalization of environmental costs. Dynamic efficiency — a firm's ability to innovate and gain economies of scale — is not only a significant influence on its ability to compete but also a principal deter-

[1] This research was conducted in partial fulfilment of the requirements for a PhD at the Science Policy Research Unit of the University of Sussex. I would like to thank Alyson Warhurst for her generous guidance, support, and encouragement throughout the work on the thesis. I am also very pleased to acknowledge the support of the International Development Research Centre (Canada), the Overseas Development Administration (United Kingdom), the Centro de Estudios Mineria y Desarrollo (Bolivia), and the Bolivian Ministry of Mines. I wish to convey my thanks to representatives of the mining firms for making information and access available and to the many individuals who provided guidance, criticism, and invaluable support during the research. I would also like to thank Gavin Bridge for his kind assistance in editing the dissertation in preparation for this chapter.

minant of its environmental performance. Increased competitiveness encourages investment in technological capability and production capacity, and this in turn reduces pollution per unit of output, whereas decreased competitiveness increases pollution per unit of output. The analysis illustrates how pollution results both from a market failure to adequately price environmental resources and from a lack of dynamic efficiency in firms. An implication for environmental policy is that regulatory initiatives to reduce pollution need also to address the dynamic inefficiency of firms, as well as considering externalities.

Theoretical and policy background

According to conventional environmental economics, some environmental degradation is an inevitable by-product of human activities. The critical issue for society is not to prevent pollution altogether but to determine the optimal level of pollution or pollution control by balancing the benefits of polluting activities against their associated costs (see, for example, Ruff 1970). According to economic theory, the independent actions of producers and consumers in competitive markets will, under certain conditions, determine these optimal levels of pollution or pollution control. For example, if the polluters and those with grievances about them are few and property rights are clearly defined, voluntary bargaining between the two parties may result in an optimal solution (Coase 1960). Real markets, however, fail to meet these conditions in several ways, and the literature on environmental economics highlights the significant role of externalities in generating excessive environmental degradation. Externalities occur when the actions of an economic agent affect (positively or negatively) the welfare of others and they are not compensated for the damage or receive benefits free of charge.

Externalities undoubtedly cause pollution levels beyond socially desirable limits but may not be the sole cause of excessive (suboptimal) environmental degradation. An emerging body of literature is therefore beginning to address the environmental effects of production inefficiencies in developing countries (Moore 1986; Barbier 1989, 1991; Pearce et al. 1990; Doeleman 1991; O'Connor 1991; O'Connor and Turnham 1991; Simonis 1992; Warhurst 1992, 1994). This literature analyzes how low levels of investment inhibit the accumulation of non-resource capital and development of organizational capabilities and skilled human resources. As a consequence, developing countries undertake production processes with lower levels of efficiency than those in industrialized countries, which results in high levels of pollution per unit of output and leads the poorest populations of the Third World to exploit the natural environment without regard to its sustainability. O'Connor (1991) and Warhurst (1994) pointed out that in the mining

industry, low educational and skills levels of workers can negatively affect productivity and the maintenance of equipment. This reduces profit and constrains a
company's capacity to invest. As a result, companies are unable to renew capital
equipment or acquire state-of-the-art equipment that pollutes less per unit of output. Similarly, a principal characteristic of the many artisanal mining operations
that prevail in developing countries is their underexploitation of ore deposits and
overexploitation of the environment's capacity to receive waste. This situation is
a result of high rates of time preference and a shortage of capital and technical
knowledge.

By suggesting that a firm's level of pollution is related to its efficiency,
this emerging literature makes a significant contribution to the analysis of excessive (suboptimal) environmental degradation. However, this analysis has two
flaws. First, although it highlights the interface between the theory of pollution
and the theory of production, it fails to systematically integrate them. Second, it
fails to develop a dynamic model of the firm as an alternative to the conventional,
static approach. This study seeks to address these shortcomings by developing and
empirically testing a dynamic model of competitiveness and environmental performance in the mining firm.

The case studies of the Bolivian mining industry illustrate how dynamic
investment in technologies and organizational forms that internalize and reduce environmental costs improve the competitive position of a company and improve its
environmental performance. The formal hypothesis underpinning the research is
that the externalization of a firm's environmental costs is determined by its competitive efficiency. Central to the analysis are two assumptions regarding the mining firm. First, competition between firms encompasses the capacity to innovate,
which enhances a firm's ability to compete through technical change. Instead of
simply maximizing their competitiveness within fixed constraints, mining firms
improve it by changing the constraints. Second, competition operates as an evolutionary process that selects between successful and unsuccessful firms. As firms
continually create and change products and modify the production process, the
social mechanism of industrial competition influences the selection of many alternative approaches to the production of goods and services. The analysis shifts the
focus from externalities to the negative environmental effects of companies facing
competitive difficulties and examines the environmental payoff for companies and
industries that can sustain and improve their competitive advantages. By demonstrating that pollution per unit of output is inversely related to a mining firm's
competitiveness, this study lends empirical support to a priori arguments that competitiveness and environmental performance converge.

Toward a dynamic theory of the mining firm

This section establishes the analytic framework for the study by developing a dynamic economic theory of the mining firm. The static approach of conventional economic theories of depletion and pollution is highlighted to show how they neglect the interrelationship between production efficiency and environmental degradation. In this section, I develop a simple dynamic model of a mining firm to show that the degree of internalization of a mining company's environmental costs depends on changes in its competitiveness. The concept of dynamic competition highlights the process by which companies compete through technical change and changes in their production factors. This model therefore draws on Porter's (1990, p. 20) distinction between static and dynamic competition:

> In a static view of competition, a nation's factors of production are fixed. Firms deploy them in the industries where they will produce the greatest return. In [dynamic] competition, the essential character is innovation and change. Instead of being limited to passively shifting resources to where the returns are the greatest, the real issue is how firms increase the returns available through new products and processes. Instead of simply maximising within fixed constraints, the question is how firms can gain competitive advantage from changing the constraints. Instead of deploying a fixed pool of factors of production, a more important issue is how firms and nations improve the quality of factors, raise the productivity with which they are utilised, and create new ones.

Competitive equilibrium and the mining firm

One of the major achievements of neoclassical economic theory has been its ability to abstract from specific differences among thousands of firms the common features underlying economic activities and the functioning of markets to explain these at a high level of generality. Competitive equilibrium, for example, is a concept based on the assumption that firms are driven by profit and select production levels to maximize their profit. Within this theoretical framework, the relationship of prices to costs is key to a firm's achieving equilibrium, as the shape or position of its cost curves will change only if the prices of the firm's production factors change.

Despite this tendency to universal abstraction, a specific theory of the mining firm and the economics of exhaustible resources has been developed to account for the concept of user cost. Hotelling (1931) established that because an ore deposit is an exhaustible resource and a unit of ore can be exploited only once, the competitive mining firm is not in equilibrium at the production level at which price equals marginal cost. Whereas in manufacturing, for instance, today's production theoretically does not limit tomorrow's, in mining, today's production very much depends on yesterday's. The maximization of profit for a mining firm must therefore take into account this opportunity cost (commonly referred to in the literature as the user cost).

The allocation of mineral production over time is therefore fundamental to the maximization of a mining firm's profits. At any point in time, a mining firm has to allocate production over time to maximize the returns from natural capital. Yet, mining companies also have to consider capital-investment programs that could increase the present value of their profits through the development of technical and organizational change. Investment can improve exploration, project development, and extraction and processing technologies. Innovation, for example, can expand the stock of exhaustible natural resources that are economically exploitable by shifting mineral resources into mineral reserves.

Mineral production and environmental externalities

Minerals production requires, in addition to ore deposits, other environmental resources, such as land, water, and air. These environmental resources provide inputs to the production process and receive its waste streams. Under free-market conditions, most of these environmental resources have no price. Consequently, the mining company never internalizes these costs, and they are imposed on society instead. But these resources contribute to society's well-being because they provide production inputs and amenity services and support life. To maximize society's welfare over time, therefore, the right balance is needed between pollution and environmental protection, which the market fails to deliver as a result of externalities related to the exploitation of environmental resources.

In Figure 1, curve B represents the marginal social benefits (MSB) of mining pollution and the demand for pollution derived from the demand for minerals. (Pollution benefits society to the extent that it derives from the production of minerals, which contribute to the welfare of society.) Its negative slope reflects that fact that for each additional unit of mineral production, society's welfare is increased less than by the last unit produced. Curve C represents the marginal social costs (MSC) of pollution. Its positive slope illustrates that for each additional unit of mining pollution, society's welfare is decreased by a greater amount than the last unit of mining pollution because fewer environmental resources are available for other purposes. The areas below curves B and C define the MSC and MSB of pollution, and at pollution level P_2 society's welfare is maximized.

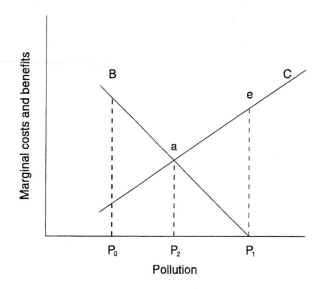

Figure 1. Optimal pollution level.

From the mining firm's standpoint, curve C coincides with the horizontal axis because pollution costs are external to the firm. Thus, mineral production will increase up to P_1. At P_1, the benefits to the mining firm are maximized. However, at these levels of pollution and production, the area P_1ae defines net social loss. The market fails to deliver P_2 because it is unable to price environmental resources and, therefore, unable to reflect the MSC of pollution (the area under curve C) adequately.

Figure 2 shows the effect on the firm of internalizing pollution costs to establish a level of environmental protection (such as set by government regulation). Under free-market conditions, X_0 is produced and there is no environmental protection. If, for example, a level of environmental protection (E_0) is required, the curve of marginal costs (which includes user costs) shifts from MC ($E = 0$) to MC ($E = E_0$). Accordingly, production is reduced from X_0 to X_1, and the firm internalizes MC given by the distance AC.

The combined effect of pollution and user cost

A number of conclusions follow from the above discussion. First, as ore deposits are exhaustible, their exploitation involves a depletion cost (user cost), which is delivered by the market and is therefore internal to the mining firm. Second, under

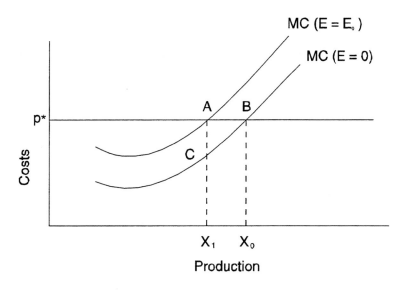

Figure 2. The firm's marginal cost (MC) of pollution.

free-market conditions, excessive pollution will take place because some environmental resources are inadequately priced by market mechanisms. Further, as production functions are constant under conditions of static competition, reductions in pollution take place only if an external agent (such as a governmental agency) encourages the mining firm to protect the environment. This lowers the mining firm's output. The mining firm's pollution costs include, therefore, both expenditures on production factors used for environmental protection and the income losses arising from a reduction in output. Third, a striking feature of this discussion is that user and pollution costs are both related to the exploitation of natural resources, although user and pollution costs have traditionally been analyzed independently.

To integrate both approaches, let us first assume, for simplicity, that firms are in equilibrium and that mineral production does not require capital but only labour. Labour can be employed either in the production of ore or in environmental protection. Thus, the marginal costs of the mining firm only include labour and user costs. Under these conditions, an increase in pollution costs resulting from changes in environmental regulations brings about two conservation effects. First, marginal reserves are changed into resources, and the total amount of ore to be exploited over a mining project's lifetime is reduced. Second, rising pollution costs make the mining firm's initial production path no longer optimal. To maximize current profits, costs must be transferred from the present to the future because future costs have a smaller present value than current costs. Thus, environmental costs introduce a conservation effect as current output decreases in preference to future output. However, an increase in the interest rate raises the rate of growth of user costs and encourages a mining firm to bring forward future production. This modifies the optimal production path by increasing current production in preference to future production and by reducing the life span of the ore deposit. The amount of current pollution increases, although the total amount of pollution remains constant over the life span of the deposit. This may harm the environment if, in periods close to the present, environmental thresholds are close to being surpassed.

The introduction of capital complicates the analysis. On one hand, the increase in the interest rate promotes the expansion of ore production in the periods close to the present. On the other hand, it also increases capital costs and prevents an increase in present production. Thus, as the mining industry is capital intensive, it is uncertain whether a change in user cost will modify both the mining firm's production and its pollution paths. This suggests that introducing user costs into the analysis does little to help us understand the mining firm's pollution costs. It would appear, therefore, that under conditions of static equilibrium (and as conventional environmental economists implicitly conclude), the problem of pollution can be adequately analyzed independently of depletion. This means that if production functions are given and firms only compete through hiring the production factors that best fit their production functions, then the problem of depletion of natural resources is essentially independent of that of environmental pollution. The following section moves the discussion forward by analyzing the relationship between user and producer costs, assuming that mining firms also compete through changing their production functions. Under such dynamic-competition conditions, a mining firm may both increase ore production and internalize environmental costs through technical change.

Dynamic competition, production capacity, and technological capability

Thus far, the analysis has considered the mining firm a simple system, with production factors organized to produce ore concentrates. Two assumptions underlie this representation. First, it is assumed that a particular firm has technical relationships linking output levels with input specifications and combinations and that changes in output involve changes in the use of production factors, according to the technical possibilities given by the mining firm's production function. Second, it is assumed that output levels, or the rates of ore extraction over time, are those that maximize the present value of the mining firm's profits. Therefore, changes in metal and production-factor prices prompt changes in output and in the use of production factors to maximize the current value of the firm's profits or to minimize the current value of its losses. Again, these changes conform to the technical possibilities given by the firm's production function.

In reality, however, firms are driven by profits, and the profits accrued by firms depend on their production functions. Thus, assuming that firms change and upgrade their production functions is more realistic than assuming that firms' production functions remain unchanged. Moreover, the many technical changes in the minerals industry in the last century indicate that competition involves innovation and that, consequently, firms relentlessly change their production functions. An analysis of the mining firm under conditions of dynamic competition is therefore empirically, as well as theoretically, justified.

The key feature of an industrial firm under conditions of dynamic competition is that it devotes its resources to producing not only final goods or services but also technical change to improve the efficiency of the production process and upgrade the firm's production function. As pointed out by Bell and Pavitt (1993), the firm has two stocks of resources: production capacity and technological capability. Technological capability incorporates the resources needed to generate and manage technical change, which is any change that modifies the levels of efficiency for a given production capacity. Technological capability includes mainly intangible assets, namely, knowledge, skills, experience, and institutional structures and linkages (in firms, such as collaboration among process-engineering departments; between firms, such as cooperation between the user and the supplier; and outside firms, such as connections with government research and development [R&D] departments). Consequently, the firm may accumulate two types of stock: production capacity and technological capability. Output and technical change are the respective outcomes of those stocks and are accumulated according to the profit-maximizing behaviour of the firm.

Because technical change seeks hitherto unachieved production results, the effects of accumulated technological capabilities are uncertain. When investment in technological capability is carried out, knowledge and experience are insufficient for one to adequately assess the overall effect of technical change. Nonetheless, technological accumulation is a key weapon in industrial competition. As Porter (1990) suggested, the fiercer the competition, the greater the incentive for companies to accumulate technological capabilities and to generate innovations. This causes firms to develop in an evolutionary fashion, which occurs for two

reasons. First, to survive, firms have to continuously adjust their stocks of production capacity and technological capability to meet changing production and
competition conditions. Accordingly, firms strive for survival by accumulating
production capacity and technological capabilities. Second, competition discrimi-
nates between the successful and unsuccessful firm strategies. Firms accrue profits
according to their degree of success. In this way, successful firms increase over
time their capacity to invest, which enables them to accumulate production capacity and technological capability, which further improves their competitiveness and
strengthens their competitive advantage (Nelson and Winter 1982).

A dynamic model of the mining firm

The following model builds on the above theoretical framework and represents an
original contribution to the study of environmental economics and exhaustible
resources.

Because investment decisions modify a firm's stocks of production capacity and technological capability, the total investment of a mining firm in period t
(I_t) can be expressed by the following equation:

$$I_t = I_t^Q + I_t^{TA}$$

[1]

where I_t^Q and I_t^{TA} are the investments in production capacity and in technological
capability in period t, respectively. In period t, the investment in production capacity depends on the gap between the optimal production capacity and the actual
production capacity in period $t - 1$ ($Q_{t-1}^{opt} - Q_{t-1}^a$); the interest rate ($r_t$); the firm's
changes in profitability, which are estimated by its changes in competitiveness
(dq_t) in period t; and other determinants represented by the variable v_t. The larger
the gap between optimal and actual production capacity in the last period, the
greater the current rate of investment. If in period $t - 1$ there was excessive production capacity, $Q_{t-1}^{opt} - Q_{t-1}^a$ becomes negative, with the result that in period t
reductions in production capacity are encouraged.

These changes in competitiveness in period t (dq_t) are a proxy for the variations in the firm's rate of return. It is assumed that if a company increases its
market share, its profitability will also increase because the competitive success
of a mining firm determines its profitability in the long run. The optimal production capacity over time is a function of the productivity levels attainable by the
firm, given the technology and the firm's mineral resources (T_t^a), the prices of
inputs used in the production process (P_t^i), the price of ore that is sold (P_{t-1}^o),
changes in competitiveness (dq_{t-1}), and other factors such as mining policies (x_{t-1})

prevalent in period $t - 1$. Consequently, in period $t - 1$, the following function describes a mining firm's optimal production capacity:

$$Q_{t-1}^{opt} = o(T_{t-1}^{a};\ P_{t-1}^{i};\ P_{t-1}^{o};\ dq_{t-1};\ x_{t-1}) \qquad [2]$$

In a given period, the mining firm's planned technological capability depends on its actual production capacity (Q_{t}^{a}), changes in its competitiveness during the same period (dq_{t-1}), and other factors (y_{t-1}). Thus, in period $t - 1$, the planned technological capability of a mining firm is described by the following function:

$$TA_{t-1}^{wi} = p(Q_{t-1}^{a};\ dq_{t-1};\ y_{t-1}) \qquad [3]$$

The effect of changes in a mining firm's competitiveness (dq_{t-1}) on its planned technological capability and optimal production capacity is the kernel of the model. If a decline occurs in the competitiveness of a mining company, changes in technological capability may conflict with variations in production capacity because of changes in anticipated user costs in relation to the planned user costs.[2] This is because a decline in competitiveness indicates that a firm cannot sustain its ability to compete and that capital losses may occur over time. In this situation, the firm minimizes its losses or maximizes its benefits by increasing its current production according to the anticipated decline in the growth rate of the user cost. However, because increased technological capability increases the profitability of future production only, the firm may reduce its investment in technological capability. Consequently, investments in production capacity may be encouraged in periods closer to the present as the firm's optimal production capacity increases, but investment in technological capability is greatly reduced as the firm finds the accumulation of technological capability less attractive.

In contrast, if a mining company is improving its competitiveness, the growth rate of its anticipated user costs increases faster than planned. An upward trajectory of a firm's ability to compete indicates that its production efficiency will be enhanced over time, and the firm can earn greater profits per tonne mined than when its competitiveness remains stable. As mineral resources can be exploited

---

[2] Planned and anticipated user costs are the user costs as perceived by a mining agent at two different moments. The planned-user-cost path is the one consistent with the optimal production path when an investment project is carried out. The anticipated-user-cost path is the one that at any point in time the mining agent anticipates will prevail in the future. Consequently, if planned- and anticipated-user-cost paths coincide, the mining firm's present production plan is optimal. However, if the anticipated-user-cost path differs from the planned-user-cost path, the production plan is then no longer optimal and has to be adjusted to maximize profits.

just once, the possibility of better exploitation of the mineral resources in the future — once production efficiency is enhanced — encourages a firm to transfer production from the present to the future. Therefore, the improvement in competitiveness of a mining firm increases its planned technological capability. This has no adverse effects on the accumulation of production capacity because of the complementary relationship between the incentives to invest in technological capability and those to invest in production capacity. Success prompts mining firms to increase their investments in exploration and the development of ore deposits and encourages them to accumulate production capacity as competitors are forced to leave the industry. This entails a steady increase in the optimal production capacity of successful firms.

The environmental-performance function

The environmental performance of a mining firm is defined by the pollution (environmental degradation) it generates per unit of output, rather than by the absolute level of pollution. It is important to consider production efficiency along with pollution. For example, if a smelter emits 100 t of SO_2 into the atmosphere in producing 1 000 t of metal/d, it emits 100 kg of SO_2/t of metal. If, however, the smelter increases production to 3 000 t of metal/d but increases SO_2 emissions to only 150 t/d, its pollution per tonne of metal produced is reduced by half. Thus, changes in a mining firm's environmental performance can be described by changes in pollution per unit of output. In period t, the pollution per unit of output of a mining firm (Θ_t) is a function of the waste produced per unit of output (W_t), the way waste is disposed of (D_t), the toxicity of the waste (τ_t), and the consumption of complementary environmental services per unit of output (C_t). Thus,

$$\Theta_t = a(W_t;\ D_t;\ \tau_t;\ C_t) \qquad [4]$$

The amount of waste per unit of output depends on the ore grade and the percentage of ore-grade dilution in the extraction. Ore grade is determined by the metal content per unit of ore, so decreases in ore grade will decrease metal recovery per unit of ore. The amount of waste per unit of metal output will thus increase as ore grade decreases. Unlike ore grade, which is a parameter given by nature, the degree of ore-grade dilution is a technical parameter, indicating the difference between ore grade and head grade (the grade of mined ore that is fed to a concentration plant). Ore-grade dilution results from imperfections in the blasting operation that increase the waste per tonne of ore extracted from the mine. Thus, if ore-grade dilution increases, the amount of waste also increases, and vice versa. Given a fixed amount of waste, the pollution per unit of output varies with the system of waste disposal and the toxicity of the waste stream. Thus, dumping mining tailings directly into the environment has a greater impact than disposing of the same tailings in special reservoirs. The treatment and reuse of mine and mineral-processing water can reduce water consumption per unit of output and decrease the amount of water pollution.

The model assumes that two factors determine waste-disposal practices at mining operations: environmental regulations (g_t^{Re}) and the ability of the environment to assimilate waste (S_t^w). In the absence of regulations requiring the internalization of environmental-damage costs, mining firms minimize costs by maximizing the use of the environment as a receiver of waste. For a given set of environmental regulations, changes in a mining firm's waste-disposal practices are related to the ability of the environment to assimilate waste. The environment's ability to assimilate waste is limited, so the greater the amount of waste produced in a given period, the more quickly the assimilation limit is reached. Therefore, changes in the ratio of firm's production capacity to the ability of the environment to receive waste (Q_t^a/S_t^w) will lead to changes in the waste-disposal practices of mining firms. Thus,

$$D_t = c(Q_t^a/S_t^w; g_t^{Re}) \qquad [5]$$

Toxicity of mining wastes is assumed to be a function of environmental regulations (g_t^{Re}) that are mainly related to environmental-quality standards; metal recovery in mineral-processing activities (σ_t); loss of reagents per unit of output (ϕ_t); and the natural properties of the ore, particularly those that determine acid mine drainage. As the metal-recovery rate decreases, more metal is discharged into the environment. Similarly, reagents lost during mineral processing can leach into groundwater, surface water, and soils. Therefore, the degree of toxicity of waste generated by a mining firm can be described by the following function:

$$\tau_t = d(\sigma_t; \phi_t; Fe_t; g_t^{Re}) \qquad [6]$$

This assumes that consumption of environmental resources per unit of output is determined by their degree of scarcity and by existing environmental regulations. Scarcity discourages consumption, and consumption increases scarcity. Natural availability and the effects of demand from a firm and from other consumers determine the availability of these resources at the mining site (S_t^o). Consequently,

$$C_t = e(Q_t^a/S_t^o; g_t^{Re}) \qquad [7]$$

Thus, in period t, changes in a mining firm's environmental performance can be described as follows:

$$d\Theta_t = f[d(Q_t^a/S_t^\circ); \; d(Q_t^a/S_t^w); \; d\delta_t; \; d\sigma_t; \; d\phi_t; \; dFe_t; \; d\Gamma_t; \; dg_t^{Re}] \qquad [8]$$

This model is consistent with the results of a conventional environmental-economics analysis because it demonstrates that under conditions of static competition, changes in a mining firm's environmental performance will result only from changes in environmental regulations. Under conditions of static competition, actual production capacity is optimal (all other things being equal), so production capacity will not change over time. Similarly, environmental properties are assumed to be stable over a firm's life span, so $d(Q_t^a/S_t^\circ)$, $d(Q_t^a/S_t^w)$, dFe_t, and $d\Gamma_t$ are therefore equal to zero. Under conditions of static competition, the firm's production function and technical parameters are given, so $d\delta_t$, $d\sigma_t$, and $d\phi_t$ are also equal to zero. Under conditions of static competition, equation [8] can therefore be simply expressed as follows:

$$d\Theta_t = g(dg_t^{Re}) \qquad [9]$$

Under conditions of dynamic competition, however, production capacity and the technology used in the production process vary along the evolutionary path of the mining firm. Thus, the variables are not equal to zero, and equation [8] can be restated as

$$d\Theta_t = h(dQ_t^a; \; d\delta_t; \; d\sigma_t; \; d\phi_t; \; dFe_t; \; d\Gamma_t; \; dg_t^{Re}) \qquad [10]$$

Technical change and environmental performance

Under conditions of dynamic competition, a mining firm has incentives to improve its environmental performance. Improvements that increase metal-recovery rates or decrease ore-grade dilution and reagent losses not only increase profits but also reduce pollution per unit of output. Thus, a mining firm's investment in technological capability favours its environmental performance. Each variable can therefore be modeled as a function of investment in technological capability. For example, changes in metal-recovery rates are a function of investments in technological capability and of changes in other factors, such as head grade and concentrate grade. Similarly, changes in reagent losses are a function of investment in technological capability and of changes in other factors, such as the complexity of the ore. Because changes in production capacity are equal to investment (either positive or negative) in production capacity, changes in environmental performance can be described by the following function:

$$d\Theta_t = h(dq_{t-1}; dq_t; r_t; P_{t-1}; W_{t-1}; d\xi_t; dg_t^{Re}; d\mu_t) \quad [11]$$

This equation describes how, in period t, changes in pollution per unit of output are a function of several factors:

- Changes in a mining firm's competitiveness (dq_{t-1}; dq_t), both in period t and in period $t - 1$;

- The interest rate in period t (r_t);

- The prices of inputs and outputs in period $t - 1$ (P_{t-1});

- The stocks accumulated by the firm until period $t - 1$ (W_{t-1}), which include the firm's production capacity and technological capability;

- Changes in some technical parameters that are caused not by technical changes but mainly by changes in the natural properties of the ore deposit ($d\xi_t$);

- Changes in environmental regulations (dg_t^{Re}); and

- Changes in residual factors ($d\mu_t$).

The key feature of equation [11] is that changes in a mining firm's pollution per unit of output correlate negatively with changes in its competitiveness. This relationship has been neglected in the literature on environmental economics because researchers have assumed static competition. In contrast, the model developed here predicts that under conditions of dynamic competition, if a mining firm's ability to compete (competitiveness) is improved, its pollution per unit of output is reduced and vice versa. This is because, other things being equal, under conditions of dynamic competition the improvement in a mining firm's competitiveness increases its investments in production capacity and technological capability. Technological capability increases as the rate of growth of user costs rises over time, the efficiency of production improves, and profits increase, giving the firm easier access to investment funds. The increase in technological capability improves the firm's environmental performance by reducing ore-grade dilution and losses of reagents and increasing metal-recovery rates.

Although it is uncertain whether reductions in production capacity will occur if the competitiveness of a mining firm declines, the downward trend in the rate of growth of user costs can lead to increases in pollution per unit of output. This is because investments in technological capability are drastically reduced, and the firm attempts to minimize natural capital losses by maximizing current output. This has a negative effect on production efficiency; consequently, ore-grade dilution and reagent losses are likely to increase, and metal-recovery rates may drop. Furthermore, as financial difficulties become more serious over time, the mining firm either has to shut down or has to scale down its operations. The latter negatively affects its waste-disposal practices and consumption of complementary environmental resources per unit of output.

It is important to stress that the validity of the relationship between changes in competitiveness and changes in environmental degradation per unit of output relies on the assumption that the technological capabilities of a mining company relate to the generation and management of incremental (as opposed to radical) technical change. This is because incremental technical change underlies the properties described in the equations and relates to improvements in the process that reduce inputs (reagents and complementary environmental services) or increase outputs (overall metal-recovery rates) over time. Radical technical change, on the other hand, might relate to the development of a new process, using, for instance, new and more toxic reagents to improve metal-recovery rates. The total effects of radical technical change on pollution per unit of output are thus uncertain. The assumption that a mining firm's technological capability generates and manages incremental technical change is based on the fact that mining is a scale-intensive industry and one in which incremental improvements of process technology dominate technical development. It must be emphasized that the dynamic model in no way suggests that technological development in mining has been protective of the environment. However, as long as incremental technical change dominates the pattern of technical change in a firm, the model demonstrates that improvements in a firm's competitiveness can result in improvements in its environmental performance and vice versa.

The interface of user and pollution costs revisited

A mining firm's investment in technological capability can modify the amount of economically exploitable ore over time and therefore can affect a mining firm's user costs. Incremental technical changes that enable a firm to economically exploit lower grade ore and increase recovery rates, for example, can reduce both

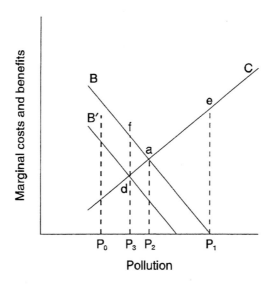

Figure 3. Excessive pollution resulting from externalities and dynamic inefficiency.

present user costs and pollution per unit of output. Thus, under conditions of dynamic competition, the theory of depletion and the theory of pollution appear to be significantly related, whereas this is not so under conditions of static competition. The overlap of theories makes it evident that excessive environmental degradation originates not only in the inability of the market to appropriately price environmental resources but also in competitive inefficiency. In other words, firms unable to accumulate technological capability and production capacity mismanage the environment, both as a source of raw materials and as a receiver of waste.

Figure 3 illustrates how excessive environmental degradation is related both to externalities and to competitive inefficiency. Curve B shows the demand for pollution derived from the demand for minerals, and curve C shows the marginal costs of pollution to society; P_2 represents the optimum pollution level under conditions of static competition. Under conditions of dynamic competition, however, improvements in production efficiency cause curve B to shift to B', as a result of incremental technical change and the enlargement of successful firms' production capacity. These improvements further reduce pollution per unit of output, from P_2 to P_3. At P_3, society's betterment increases by the amount given by the area *afd*. Conversely, if the pollution level is at P_1, society's excessive pollution level — $P_1 P_3$ — comprises two parts:

- P_1P_2 is the level of suboptimal environmental degradation resulting from externalities associated with the use of environmental resources; and

- P_2P_3 is the excessive level of environmental degradation resulting from dynamic inefficiency, which discourages reductions in the mining firm's pollution per unit of output.

Testing the model

The main prediction of the model — that, *ceteris paribus*, pollution per unit of output is inversely related to a mining firm's competitiveness — was evaluated in the context of the Bolivian mining industry. The Bolivian mining industry provides a good case study because it satisfies two conditions for a simple empirical test of this study's hypothesis:

- The absence, in practical terms, of a systematic framework of environmental regulations relating specifically to mining; and

- Clear signs that the structure of the mining industry has dramatically changed since the 1980s, after a period of relative stability during the 1960s and 1970s.

Thus, the main effects of a company's investment decisions on its competitiveness and environmental performance are likely to emerge from the analysis.

Structure of the Bolivian mining industry

Bolivian mineral producers can be classified into three types, based on their ownership structure and size:

- The state-owned mining company, Corporación Minera de Bolivia (COMIBOL), which comprises all mining operations nationalized in 1952 and has been for more than three decades the largest single mining company in Bolivia;

- Privately owned large- and medium-scale mining operations; and

- Privately owned small-scale mining operations and mining cooperatives, which are typically labour intensive.

Table 1. Bolivia's zinc, tin, silver, and gold production, 1960–92.

	Share of national production (%)		
	State mining (COMIBOL)	Large- and medium-scale mining	Small-scale mining
Zinc			
1960	80	15	5
1971	69	—	31
1980	60	33	7
1990	24	61	15
Tin			
1960	65	12	23
1970	64	22	14
1980	68	22	10
1990	34	11	55
Silver			
1960	85	3	12
1970	81	4	15
1980	84	11	5
1990	36	48	16
Gold			
1960	—	48	52
1970	2	30	68
1980	—	21	79
1992	1	90	9

Note: COMIBOL, Corporación Minera de Bolivia.

In 1990–92, zinc, tin, silver, and gold production represented 90% of total nonferrous-minerals production by value, with antimony, bismuth, copper, lead, and wolframite (tungsten) accounting for the remainder.

Table 1 shows how the structure of Bolivia's zinc, tin, silver, and gold production changed between 1960 and 1992. COMIBOL was the most significant producer of zinc, tin, and silver until the early 1980s, but its importance declined by 1990 as the private mining operations increased their share of national production. Because the structure of the Bolivian mining industry changed dramatically during the 1980s, one might expect to find significant changes in both the competitiveness of the companies and in the ways they carry out production. The Bolivian mining industry is therefore a suitable case for testing whether, in the absence of environmental regulations, a mining firm's trajectory of pollution per unit of output will correlate inversely with its trajectory of competitiveness.

Indicators of a company's environmental performance

The environmental performance of a mining company was assessed by reference to the firm's pollution per unit of output. Pollution was evaluated by changes in the following parameters: metal-recovery rates; reagent consumption; and solid- and liquid-waste-disposal practices at the mining site.

Metal recovery and head grade

Mineral production involves the processing of large amounts of ore to obtain small amounts of metals. Because 10% of the metal content cannot be economically recovered, a proportion of the metal content is disposed of in the environment. The ratio of metal recovered to total content in the feed ore is the metal-recovery rate, expressed as a percentage.

Water recycling and treatment

Water is a very important input in the mining industry. All mining operations in Bolivia use water to process minerals, and most of these operations take place in areas where water is scarce. Cations of heavy metals and chemical substances used in the dressing process often contaminate mine and mineral-processing water. The pollution caused by a mining operation is, therefore, heavily affected by its water- and waste-management practices.

Disposal of waste

Mining operations generate a variety of solid wastes. Waste (barren rock, coarse material, overburden, tailings) comes out of mines and ore-dressing plants; and scrap comes from mining and mineral-processing equipment. Because pollution per unit of output relates not only to the quantity and toxicity of the waste but also to waste-disposal practices, three categories of disposal were considered:

- The absence of disposal systems, or the unrestricted use of the environment's sink function;

- The disposal of waste but failure to undertake reclamation at the closure of the operation; and

- Waste disposal and reclamation.

Reagent consumption

Thiosalts and other chemical substances (such as sulfuric acid, sodium cyanide, sodium hydroxide, zinc sulfate, diesel oil, amine, and pine oil) are used in processing minerals and are ultimately discharged into the environment. Because water and soils at Bolivian mining operations are not systematically analyzed, I estimated the loss of reagents in tailings from reagent consumption per tonne treated.

Trajectories of competitiveness

This section presents the results from empirical analysis of the relationship between competitiveness and pollution per unit of output for selected mining companies. I used least-squares regression analysis and analysis of variance (ANOVA) to distinguish periods in which a company's competitiveness remained constant from those in which competitiveness either rose or fell. I then matched these trajectories of competitiveness to trajectories of pollution per unit of output to ascertain whether there was a significant correlation between competitiveness and environmental performance.

Empresa Minera Inti Raymi

Empresa Minera Inti Raymi exploits gold and silver in its Kori Kollo deposit in Oruro, about 200 km from La Paz. Although Kori Kollo has been exploited since colonial times, its development was limited until the 1980s because production of sulfide concentrates was not commercially feasible. However, massive oxidized and sulfide gold–silver deposits were discovered in the early 1980s, and in 1982 Empresa Minera Inti Raymi was founded to exploit these deposits. To exploit the oxidized deposit, Inti Raymi successfully introduced a heap-leaching operation, which expanded from 400 t/d to 4 000 t/d by 1987. A 14 500 t/d agitation-leaching project was initiated In 1993 to exploit the sulfide deposit. Figure 4 illustrates that Inti Raymi has followed an upward trajectory of competitiveness. Whereas world gold production increased 58% between 1984 and 1992, Inti Raymi's market share rose by 2 150%, from 0.004 to 0.090.

Corporación Minera de Bolivia

COMIBOL was established to exploit mining concessions belonging to Patiño, Hochschild, and Aramayo, three corporations that were nationalized in 1952. Until the middle of the 1980s, COMIBOL underwent few changes. As part of the structural adjustment of Bolivia's economy that began in 1986, COMIBOL was gradually restructured: some of its operations were transferred to cooperatives, and

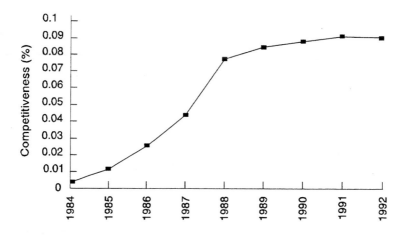

Figure 4. Inti Raymi's trajectory of competitiveness in gold.

COMIBOL actively sought joint-venture partners for other operations. Although minerals continued being significant to the Bolivian export economy (44% in 1990), COMIBOL's contribution decreased from 70% in the 1950s to around 10% by 1990. For 1970–92, COMIBOL's trajectories of competitiveness in tin, zinc, and silver were fairly similar.

Figure 5 illustrates the case for tin. Although competitiveness increased in 1970–77, it fell severely in 1978–87 in response to falling commodity prices. In 1988–92, however, COMIBOL's competitiveness in tin, zinc, and silver began to improve, although its market-share levels were lower than in the 1970s.

I took the trajectories of competitiveness and broke these down into sub-trajectories of increasing or decreasing trends in competitiveness. I then used least-squares regressions to evaluate these trends and used an ANOVA to distinguish periods in which competitiveness remained statistically constant from those in which competitiveness statistically rose or fell. The analysis showed that COMIBOL's competitiveness in tin, zinc, and silver production decreased significantly in 1978–87.

I evaluated COMIBOL's environmental performance at three mining operations: Catavi, Colquiri, and Unificada. Until the tin crisis in 1987, Catavi was COMIBOL's largest tin operation, and all COMIBOL interviewees regarded Catavi as COMIBOL's best-organized mine. I selected Catavi for analysis because it best represents COMIBOL's severe decline in competitiveness in tin. Colquiri is a major tin–zinc operation that improved its mining and ore-dressing operations

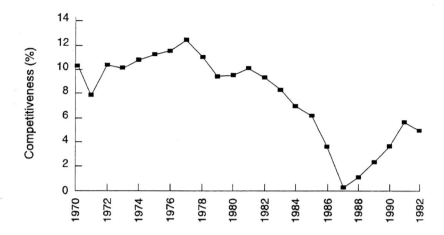

Figure 5. COMIBOL's trajectory of competitiveness in tin.

in 1988 and 1989 by introducing sublevel stoping and redesigning the concentrator. I selected Colquiri for analysis not only because it is COMIBOL's largest zinc producer but also because it made considerable efforts to improve the efficiency of its operations. Unificada, a significant producer of tin, silver, and zinc, illustrates COMIBOL's attempt to diversify production away from tin. Unificada is also the location of one of COMIBOL's major technical changes — tin volatilization. The failure to manage this technical change was one of the main causes for COMIBOL's decline in competitiveness in tin.

Compañía Minera del Sur

Compañía Minera del Sur (COMSUR) is a composite mining corporation, comprising a number of companies and operating sites. Founded in 1968, it enlarged and diversified its tin operations in the 1970s and 1980s, establishing a basis for becoming Bolivia's most important producer of zinc and lead in the 1990s. Zinc accounts for 75% of total production value, and in 1990 Rio Tinto Zinc acquired 30% of the company's equity.

The trajectories of competitiveness for COMSUR's zinc and silver production appear to have two general trends: no significant change (in 1974–82, for zinc, and in 1977–85, for silver) and a steady rise in competitiveness (in 1982–92, for zinc, and in 1985–90, for silver). Figures 6 and 7 illustrate the remarkable improvement (more than 300%) in COMSUR's competitiveness in zinc and silver, respectively.

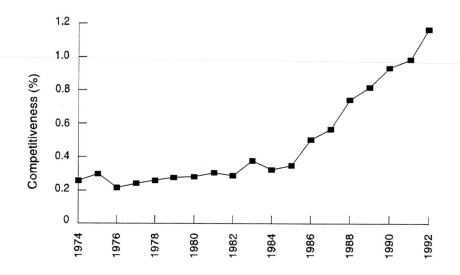

Figure 6. COMSUR's trajectory of competitiveness in zinc.

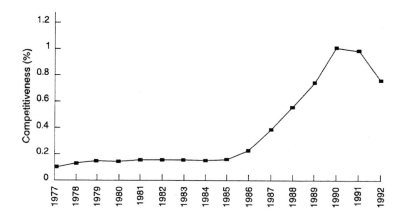

Figure 7. COMSUR's trajectory of competitiveness in silver.

An ANOVA of COMSUR's competitiveness in zinc and silver showed that the increases in competitiveness in 1982–92 for zinc and in 1985–90 for silver were statistically significant. In the case of COMSUR (as well as in the case of Inti Raymi), only periods of improved competitiveness were analyzed in relation to environmental performance. I selected the mining operations of Porco and COMCO for this. Although COMSUR's upward trajectories of competitiveness in zinc and silver may have been related to COMCO's development (at the end of the 1980s) and the acquisition and sustained production growth of Caballo Blanco S.A. (bought in 1980) and Quioma S.A. (bought in 1986), the case of Porco (COMSUR's leading mining operation) encapsulates COMSUR's success. Until the early 1980s, Porco was a fairly small operation (350 t/d), but by 1992 its zinc production had risen by 150%. In 1992, an enlargement project, from 800 t/d to 1 200 t/d, was completed, and Porco is now Bolivia's largest zinc mine (1 200 t/d). COMCO is a 1 000 t/d heap-leaching silver operation, reprocessing old oxidized tailings since 1989. COMCO may represent a turning point in COM-SUR's development: technical operations represent a departure from the traditional production methods of the Bolivian mining industry (gravimetric concentration and flotation), and COMCO is the only project that COMSUR developed from scratch.

Central Local de Cooperativas Mineras Cangallí

When the state took over the largest mining groups in 1952, mining concessions in the Tipuani–Tora region were nationalized and divided into 10 sectors, with 2 200 mining concessions. The cooperatives located in the 9th sector established the Consejo de Cooperativas Cangallí, which in 1982 became the Central Local de Cooperativas Mineras Cangallí (CECOCA). CECOCA is the legal owner of more than 800 mining concessions, distributed among 12 cooperatives, and each cooperative freely controls the concessions it has. Although CECOCA coordinates activities that require the cooperatives to collaborate (such as negotiating an agreement to drag the Tipuani river) and provides technical and financial support to the cooperatives, it neither formulates corporate policies nor interferes in any individual cooperative's business.

I selected one of CECOCA's cooperatives — the Cooperativa Aurífera Rosario California Ltda — for detailed study because it pioneered the major technical changes (surface mining and mechanization) that were diffused throughout the region during the early 1980s and is one of the gold cooperatives that has achieved the highest degree of economic success.

Figure 8 shows Rosario California's "shadow trajectory" of competitiveness in gold. Most of CECOCA's cooperatives make no systematic records of production parameters, not even of output. The shadow trajectory was therefore calculated using data from the Federación Regional de Cooperativas Auríferas (regional federation of gold cooperatives), which has since 1982 estimated the cooperatives' total gold production on the basis of archives and oral reports. In a period in which world gold production rose by 80%, Rosario California's shadow trajectory of competitiveness increased in 1978–86 but decreasedin 1986–91. An ANOVA showed that both the increase and the decrease were statistically significant.

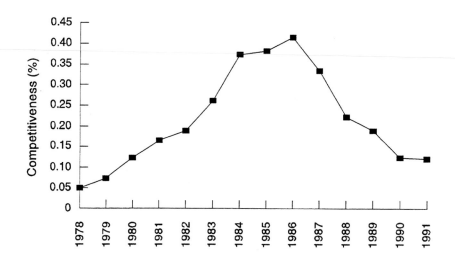

Figure 8. CECOCA's shadow trajectory of competitiveness in Cooperativa
Aurífera Rosario California.

Table 2. Trajectories of competitiveness of the four study firms, 1977–92.

Competitiveness	Firm	Mineral	Period	Operation
Decline	COMIBOL	Tin	1977–92	Catavi (Sn)
		Zinc	1978–87	Colquiri (Sn, Zn)
		Silver	1978–92	Unificada (Sn, Zn, Ag)
	CECOCA	Gold	1986–91	Rosario California (Au)
Improvement	Inti Raymi	Gold	1984–92	Kori Kollo (Au)
	CECOCA	Gold	1978–86	Rosario California (Au)
	COMSUR	Zinc	1982–92	Porco (Zn, Ag, Pb)
		Silver	1985–92	COMCO (Ag)

Note: CECOCA, Central Local de Cooperativas Cangallí; COMIBOL,
Corporación Minera de Bolivia; COMSUR, Compañía Minera del Sur.

Summary

The trajectories of competitiveness for the four study firms are summarized in
Table 2, which includes the mining operations at which the firms' environmental
performance was assessed.

Comparison of environmental performance

I divided the multiple-case study into two sets of mining operations by testing for a negative correlation between a firm's competitiveness and its pollution per unit of output. One set comprised operations for which the hypothesis of a negative correlation was not rejected; the other set comprised operations for which either the hypothesis was rejected or a correlation (positive or negative) was indeterminate. The criteria for deriving the trajectories of pollution per unit of output were based on the following:

- Variation by more than 10% in quantitative variables, such as metal-recovery rates and reagent consumption, would be deemed significant; variation of 0–10%, less significant.

- Changes in metal-recovery rates would be adjusted according to changes in head grade, based on the assumption that significant increases and decreases in head grade are equal to less significant decreases and increases, respectively, in recovery rates. If there were significant increases in metal-recovery rates and head grade, the total increase in metal-recovery rates would therefore be regarded as less significant.

- Changes in qualitative variables would be considered significant only if the firm had made major changes in water management or waste disposal (for example, the introduction of water-recycling or reclamation programs).

- Variations in pollution per unit of output due to changes (variables) in recovery rates, in reagent consumption, in water consumption, and in waste disposal would be scored to obtain aggregate indicators of changes in a mining operation's pollution per unit of output. Significant and less significant variations in a variable would score 2 and 1, respectively; no variation and indeterminate changes would score 0. A score of 4 would mean that an operation had experienced less significant changes in all four variables, so an operation scoring more than 4 would be considered as having made significant changes in pollution per unit of output. Scores of 1–4 would be considered less significant. Operations scoring 0 would be considered as having made no significant changes. The relative nature of an operation's overall score would be emphasized: if two operations scored 5 and 7, respectively, all that could be said about their environmental performance is that both had undergone significant variations, and it would be incorrect to conclude that the operation scoring 7 performed better than the one scoring 5.

Correlations of changes in competitiveness with changes in pollution per unit of output

Table 3 shows the relationship between changes in competitiveness and changes in pollution per unit of output for the eight operations examined. In only one case did the changes in a mining company's competitiveness not correlate with changes in its pollution per unit of output (Rosario California, when CECOCA's competitiveness improved). In six cases, changes in competitiveness correlated negatively with changes in pollution per unit of output as predicted by the model, and in five of these (Inti Raymi, Porco [COMSUR], Rosario California [when CECOCA's competitiveness decreased], Catavi [COMIBOL], and Unificada [COMIBOL]), the negative correlation was significant. Moreover, in all these operations, changes in pollution per unit of output due to variations in recovery rates, reagent consumption, water consumption, and waste disposal had the same direction. This initial evidence supports the model's prediction that these variables will follow similar trends because changes in competitiveness reflect complementary changes in technological capability and production capacity.

In the cases of COMCO (COMSUR) and Colquiri (COMIBOL), some variables followed opposing trends, calling into question the causal linkage suggested by the model. Moreover, because of the contradictory trends in reagent consumption, water consumption, and waste disposal, the direction of COMIBOL's trajectory of pollution per unit of output at the Colquiri operation was uncertain. I therefore assessed the robustness of the model by its ability to explain not only the contradictory trends in Colquiri and COMCO but also the main reasons underlying the lack of correlation between changes in CECOCA's competitiveness and changes in pollution per unit of output at the Rosario California cooperative (1978–86).

Environmental advantages from technical changes to enhance the production process

The data in Table 3 were used to examine the role of incremental technical change in enhancing a company's competitiveness and environmental performance and the relationship between technical change, production capacity, and environmental performance.

Table 3. The relationship between firms' competitiveness and their pollution per unit of output.

| | COMSUR | | | CECOCA | | COMIBOL | | | | |
| | Inti Raymi | Porco | | Comco | Rosario California | | Catavi | Unificada | | Colquiri | |
	Au	Zn	Ag	Ag	Au	Au	Sn	Sn	Ag-Zn	Sn	Zn
Competitiveness	84–92 I	82–92 I	85–92 I	85–92 I	78–86 I	86–91 D	77–92 D	77–92 D	78–92 D	77–92 D	78–87 D
Pollution per unit of output	D(S)	D(S)	D(S)	D(L)	k	I(S)	I(S)	I(S)	I(S)	I(S)	?
Recovery rates	86–92 D(S)	88–92 D(L)	88–92 D(L)	89–92 k	80–86 k	86–91 I(?)	77–85 ?	77–85 I(L)	89–92 k	77–92 ?	78–85 k
Reagent consumption	88–92 D(S)	88–92 D(S)	88–92 D(S)	89–92 D(S)		86–91 I(S)	88–92 I(?)	77–85 NA	89–92 I(S)	90–93 D(L)	90–93 D(L)
Water consumption	86–92 k	88–92 D(L)	88–92 D(L)	89–92 k	80–86 k	86–91 k	77–92 I(S)	77–92 I(S)	77–92 I(S)	77–92 D(L)	77–92 D(L)
Waste disposal	87–93 D(S)	88–92 D(S)	88–92 D(S)	89–92 I(L)	80–86 k	86–91 I(S)	77–92 I(S)	77–92 I(S)	77–92 I(S)	77–92 I(S)	77–92 I(S)
Correlation	–(S)	–(S)	–(S)	–(L)	0	–(S)	–(S)	–(S)	–(S)	?	?
Hypothesis rejected	No(S)	No(S)	No(S)	No(L)	Yes(S)	No(S)	No(S)	No(S)	No(S)	?	?

Note: CECOCA, Central Local de Cooperativas Mineras Cangalli; COMIBOL, Corporación Minera de Bolivia; COMSUR, Compañía Minera del Sur. D, decrease; I, increase; L, large; S, small.

Table 4 summarizes the technological performance of Inti Raymi and COMSUR. Improved metal-recovery rates and reagent consumption indicate a link between environmental performance and technological performance at these sites. For example, sodium cyanide (NaCN) consumption fell by 29% at the COMCO heap-leaching operation as a result of systematic control and monitoring. The activities of a small R&D department at the Inti Raymi leaching project resulted in even more dramatic reductions in NaCN (73%) and zinc (79%) per unit treated. The COMCO and Inti Raymi operations highlight the important fact that in leaching operations, economic and environmental considerations favour the careful management of cyanide. As an executive of Inti Raymi stated during an interview, "it will be the end of the operation if a disaster takes place because of the mismanagement of the cyanide solution." Furthermore, the profitability of the project depends very much on minimizing solution losses, not only because of the cost of cyanide but also because of the fact that metal is lost, as well as the solution.

COMSUR also achieved significant reductions in reagent consumption per tonne treated at Porco, although the operation uses different mining and mineral processing methods. In 1988–92, the Porco operation reduced its consumption of complex cyanide, $CuSO_4$, Dowfroth 1014, lime, and SF 114 by 21, 22, 33, 37, and 22%, respectively, per tonne treated. These reductions resulted from new technology and expertise in process engineering. For example, COMSUR redesigned Porco's mining operations and constructed a new and enlarged (from 800 t/d to 1 200 t/d) zinc–silver and lead-flotation plant in 1988–92. COMSUR also installed equipment embodying new technology, such as a semiautogenous grinding mill, column-cell cleaning of concentrates, and computer-based process control. In addition to the modification of the original design (from lead–zinc differential float to bulk flotation for zinc, lead, and silver), these changes made the flotation process much more efficient. The optimization of variables — such as pH levels, liquid–solids ratio, and the feed of reagents — brought about reductions in reagent consumption and increases in the grade of zinc and lead concentrates. Increasing the lead-recovery rates by 11% also significantly reduced the amount of lead per unit of output discharged into the environment.

So far, an important result has emerged: the optimization of the production process through incremental technical change has a beneficial effect on both a mining company's ability to compete and its level of pollution per unit of output. In comparison with Porco and Inti Raymi, COMCO made only a modest improvement in environmental performance, because there was less technological dynamism at that site. COMCO's ore reserves did not increase over time (in contrast

Table 4. Technological performance of Inti Raymi, Porco, and COMCO.

	Inti Raymi	COMSUR	
		Porco	COMCO
Successful radical technical changes	• Oxidized deposit: open pit and heap leaching • Sulfide deposit: agitation leaching		• Oxidized tailings: heap leaching
Successful incremental technical changes	• Induced polarization and magnetic prospecting in exploration • Improvement of the heap design • Reduction of the size of mineral grain • Improvement of the watering of heaps • Optimization of the cycle and the grade of the gold-bearing solution	• Change in explosives • Column-cell cleaning • Automatic computer-based process control • Semi-autogenous grinding • Bulk flotation, instead of lead–zinc-differential float	• Redesign of the crushing and comminution circuit
Unsuccessful incremental technical changes	• Reduction in the evaporation of the gold-bearing solution		• Replacement of dry- with wet-grinding

Note: COMSUR, Compañía Mineral del Sur.

to those at Porco and Inti Raymi), and because of this shortage of reserves, COM-SUR was unable to commit itself to changing from dry grinding to wet grinding, which would have significantly reduced its costs. In addition, political opposition to COMCO from the civic institutions of Potosi magnified the long-term risks of this project.

Thus, the analysis shows that technological performance is linked to competitiveness and environmental performance — the discharge of wastes such as heavy metals and reagents per unit of output decreases with improvements in production efficiency. The output–waste ratio, just like the output–input ratio, is an indicator of the production efficiency of an industrial operation. Improvements in the production efficiency of a mining company result in more efficient use of the environment and involve greater internalization of pollution costs.

Misuse of environmental resources due to unsuccessful technical change

Increased pollution per unit of output is also related to the mismanagement of technical change (innovation inefficiency). On one hand, because of unsuccessful technical change, the decline in metal-recovery rates cannot be arrested, and the waste generated per unit of output increases. On the other hand, unsuccessful technical change has negative effects on a company's competitiveness. This brings about significant reductions in production capacity, which result in greater externalization of costs to the environment. COMIBOL's Catavi and Unificada operations and CECOCA's Rosario California cooperative (1986–91) illustrate these processes well.

CECOCA's Rosario California

By the second half of the 1980s, sands bearing coarse gold at the Tipuani river were depleted. CECOCA's cooperatives approached this problem in two ways. First, they developed new operations downstream, at the Kaka river. Second, they altered the course of the Tipuani river to exploit the sand that had settled on the bedrock (Table 5).

The sands at the Kaka river bear significant amounts of fine gold, and mercury was used for gold recovery. However, to recover the mercury from the amalgam, CECOCA used rudimentary techniques. This brought about a significant discharge of mercury vapour into the atmosphere and increasing losses in efficiency in reusing the mercury. Upstream, CECOCA's first attempts to alter the course of the Tipuani river were successful because they were carried out where the riverbank is quite wide. However, for subsequent work where the Tipuani river

Table 5. Unsuccessful technical changes at Catavi, Unificada, and Rosario California.

COMIBOL		CECOCA
Catavi	Unificada	Rosario California
1970–85 • Cassiterite flotation (fine tin tailings) 1987 to present • Exploitation of mill tailings	1970–85 • Tin volatilization 1987 to present • Diversification from tin to zinc–silver	1986–91 • Channeling of the river • Mercury and gold amalgamating • Exploration of the old river bed

Note: CECOCA, Central Local de Cooperativas Mineras Cangallí; COMIBOL, Corporación Minera de Bolivia.

enters a very narrow valley between steep hills, channeling the river would require specialized knowledge of hydraulics, rock mechanics, and other branches of engineering to channel the river appropriately, knowledge that had not been accumulated by CECOCA or the Rosario California cooperative. Thus, at Huarachani, Rosario California made unsuccessful attempts to alter the course of the Tipuani river. Landslides could not be prevented, and the mining activities originally planned to occur over 4 years could only partially succeed for about 6 months. This had a devastating effect on Rosario California's income and competitiveness. From 1988 to 1992, production capacity fell by 66%, from 2 400 t/d to 800 t/d. Furthermore, because of the failure to channel the river, the increase in waste per unit of output discharged into the river was up to six times higher than anticipated. Between 1986 and 1991, pollution increased significantly at Rosario California, as the operation produced greater amounts of barren rock and toxic wastes (mercury) per kilogram of gold.

COMIBOL's Catavi and Unificada

At its Catavi and Unificada operations, COMIBOL introduced cassiterite flotation and tin volatilization between 1978 and 1985 in an effort to deal with a long-term decline in tin-recovery rates. From 1977 to 1985, tin head grade fell by 39% at Catavi and by 26% at Unificada. Significant increases in recovery levels can be obtained by using gravimetric and flotation methods to produce low-grade concentrates (assaying Sn at up to 5%), which the volatilization process can subsequently upgrade. In the 1960s, COMIBOL embarked on the acquisition and development of cassiterite-flotation and tin-volatilization technologies (Garret 1968). By the early 1970s, the Instituto de Investigaciones Minero Metalúrgico (IIMM, institute for the study of mining and metallurgy), an R&D institution supporting Bolivian mining and mineral-processing activities, had developed an industrial cassiterite-flotation process specifically for Bolivian ores.

The introduction of cassiterite flotation into COMIBOL's operations failed because of delays and a lack of capability to optimize the process at the plant site. COMIBOL established an industrial cassiterite-flotation plant at Colquiri in 1979, about 2.5 years after its original deadline and 8 years after it installed a 100-t/d semi-industrial plant. COMIBOL was unable to improve or even maintain the contribution of the cassiterite-flotation plant to total recovery rate. For example, after a peak in production (1977) at the El Kenko cassiterite plant at Catavi, the amount of ore treated, the recovery rate, and the concentrate grade greatly decreased, declining from 14% of Catavi's total production in 1977 to 2% by 1983.

However, COMIBOL's inability to formulate and introduce coherent plans adapted to the specific technical problems of its tin operations prevented it from fully exploiting the advantages of this technical change in Catavi. From 1977 on, the underground high-grade reserves in the Catavi mine were severely reduced, and surface low-grade reserves from old tailings increased in importance. The tin content of these surface resources was similar to that of the ore mined and fed to the sink-and-float plant. Because stripping, blasting, and crushing are unnecessary for exploiting old tailings, COMIBOL could have significantly reduced its costs if it had had the technical capacity to process these surface resources. Further, this would have provided an opportunity to reclaim tin tailings and greatly reduce acid mine drainage. Cassiterite flotation has been viewed as a promising technology for reprocessing the millions of tonnes of ore accumulated in Bolivian tin tailings. However, the Catavi operation had none of the organizational and technological capabilities required to manage cassiterite flotation in the context of a radical transformation from an underground to a surface operation, so COMIBOL was unable to change these low-grade reserves into economically exploitable resources.

COMIBOL's lack of ability to manage technical change also explains its failed attempt to introduce tin-volatilization technology at Unificada. This case, however, highlights the key role of management, rather than technical ability. COMIBOL undertook three major activities to introduce tin-volatilization at La Palca (Unificada): managing the user–supplier relationship in the context of a turnkey contract with a Russian firm, Machino Export; locating the plant and building the plant base; and developing mineral resources to supply low-grade concentrates to the plant.

MANAGING THE USER–SUPPLIER RELATIONSHIP — The investment at La Palca suffered badly because of deficiencies in the relationship between COMIBOL and Machino Export. In particular, the original contract had to be supplemented with covenants for the procurement of parts and equipment that had been overlooked. Almost 2 years after it signed the contract, COMIBOL realized that the 242-t/d plant was unprofitable, so it increased the plant capacity, first to 350 t/d and then to 400 t/d. By 1980, La Palca had been delayed by 6 years and investment expenses had increased from 25 to 68 million United States dollars [USD] (Canelas 1981).

LOCATING THE PLANT — Originally, COMIBOL intended to locate Unificada's volatilization plant at the site of an old volatilization plant (Taitón), built in the 1940s. However, 3 or 4 years into development, COMIBOL found the site inadequate because of a limited water supply and changed the location from Taitón to La Palca, instead of attempting to increase the supply of water. Although La Palca has an abundant supply of water, the local geological structure cannot support a 400-t/d volatilization plant. This was identified by COMIBOL's geological department, but COMIBOL was unable to assess the time and cost trade-offs involved in moving the plant to La Palca and fortifying the geological structures, rather than maintaining the plant at Taitón and undertaking work to increase the water supply (Canelas 1981).

DEVELOPING THE MINERAL RESOURCES — COMIBOL was unable to ensure an adequate supply of tin concentrates from Unificada for the volatilization plant. After 9 years' delay, La Palca started its operations in 1983. Utilized capacity of the operation averaged only 50% in the mid-1980s. Although Unificada produced low-grade concentrates for La Palca, it could supply only 35% of La Palca's capacity, at best, and because of the difficulty in acquiring low-grade concentrates from other sources after the collapse of tin prices on the London Metal Exchange, COMIBOL closed La Palca at the end of 1985. Thus, as a result of COMIBOL's mismanagement of cassiterite flotation and tin volatilization, it was unable to arrest the decrease in tin-recovery rates, and this increased the discharge of metals into the environment per tonne of treated material (see Table 3). At Catavi, tin-recovery rate fell by 13% between 1977 and 1985. At Unificada, the recovery rate decreased by 9% between 1977 and 1980, but it rose in 1981 upon completion of the La Palca volatilization plant. However, COMIBOL was unable to sustain this improvement, and the recovery rate fell again between 1981 and 1985. In addition, because COMIBOL did not make the required technical changes to feasibly exploit its low-grade reserves, the potential of its natural-resource base went unrealized during the period of advantageous market conditions in the 1970s and the first half of the 1980s. As a result of innovative inefficiency, COMIBOL actually increased its pollution per unit of output and wasted a great opportunity to reduce contamination through reprocessing and reclaiming old tin tailings.

The effects of changes in production capacity

The weakening of a mining firm's competitiveness may further increase pollution per unit of output because it encourages the firm to reduce its production capacity. This adjustment process is well represented by COMIBOL's environmental performance in 1986–92. After 1986, a fall in tin prices and the reduction of public credit to state enterprises, as part of a structural-adjustment program, put strong pressure on COMIBOL to adjust its operations.

Following its failure to manage cassiterite flotation and tin volatilization, COMIBOL adopted the short-term measure of selective mining in a desperate attempt to survive its crisis of competitiveness. Selective mining is the exploitation of only the highest-grade block reserves or the richest part of the mineral vein. This practice boosts productivity and income in the short term but reduces a mining operation's life span.

When operations were scaled down and employment fell, there was a significant externalization of environmental costs (see Jordan and Warhurst 1992). For example, with scaled-down operations, water-recycling and waste-management practices became redundant. Until 1985, the Catavi and Unificada operations disposed of their tailings in impoundments, from which water was recovered for processing. This internalization of environmental costs was justified by the volume of material handled. This put a premium on water supply, and the physical limitations of rivers and streams for disposal were exceeded. In scaling down operations (from 5 000 t/d to 360 t/d at Catavi and from 1 500 t/d to 350 t/d at Unificada), water demand was reduced (by up to 95%), eliminating the need for water recycling. Lower production levels reduced the risk of blocking the nearby Vetilla and Pailaviri river systems, so tailings were discharged into the rivers. As water recycling was no longer required, the costs of COMIBOL's adjustment to the new competitive conditions were transferred to the natural environment: reduced investment and operational costs were exchanged for greater environmental costs.

COMIBOL's crisis of competitiveness, in addition to its direct environmental effects, caused the firm to dismiss more than 90% of its work force at Catavi. Although Bolivia offers no unemployment benefits, the social cost of this adjustment was partially alleviated by the formation of cooperatives, with ex-COMIBOL employees, to exploit the Catavi mine. About 10 000 people work in the Catavi cooperatives, using artisanal methods. These artisanal operations have destroyed the infrastructure needed to recycle industrial water, and freshwater is now contaminated by chemical reagents, such as xanthates, sulfuric acid, and frothers, and is discharged, along with tailings, directly into the Vetilla river. Downstream, the water is heavily polluted and highly acidic: the pH fluctuates between 2.9 and 3.0 (Empresa Minera Catavi 1993).

Table 6. Scale of Operation at Inti Raymi, Porco, and COMCO.

		COMSUR			
		Porco		COMCO	
Inti Raymi					
(Years)	(t/d)	(Years)	(t/d)	(Years)	(t/d)
1985	400	1974–84	350	1989–present	1 000
1986	1 200	1985–90	800		
1987–92	4 000	1992–present	1 200		
1993	14 500				

Source: For Inti Raymi, Peró et al. (1992).
Note: COMSUR, Compañía Minera del Sur.

Thus far, the evidence presented (from COMIBOL in particular) provides strong support for the causal linkages postulated by the dynamic model of the mining firm. A decline in a mining firm's competitiveness causes an increase in its pollution per unit of output because it affects a company's investment expenses and, therefore, its stock of technological capability and production capacity. The Catavi case also illustrates that decreases in production capacity reduce not only a firm's nonnatural capital but also its labour force. In this situation, unemployed miners and their families have no choice but to resort to selective and artisan mining practices, which lowers production costs, but at the expense of incremental increases in environmental costs.

Internalization of environmental costs

Let us now return to the environmental performance of Inti Raymi and COMSUR and address the mechanisms by which an increase in competitiveness leads to the accumulation of production capacity, resulting in reductions in pollution per unit of output as water and waste management are improved.

The increase in production capacity at Inti Raymi and COMSUR's Porco and COMCO operations is summarized in Table 6. The increase in production capacity at a time of increased competitiveness reflects both an increased capacity to invest and a reduction in a firm's risk premium to the point at which the firm can make investments with uncertain returns.

Between 1987 and 1992, Porco's proven and probable reserves rose by 60 and 40%, respectively, as a result of systematic exploratory work to develop new ore veins at Las Santas. This enabled COMSUR not only to continue Porco's mining activities but also to enlarge its operations, from 800 t/d to 1200 t/d. While this expansion was happening, RTZ acquired 30% of COMSUR's equity. This move reflected RTZ's interest in Bolivia's mining potential and COMSUR's interest in having a partnership with a multinational to increase its own capacity to invest. The relationship between COMSUR and RTZ has had positive effects on COMSUR's environmental performance. In the case of Porco's expansion, for example, COMSUR replaced the old dressing plant with one embodying new, more efficient technology (Sutill 1993). This enabled COMSUR to reduce Porco's consumption of water per tonne treated by up to 13%. Whereas tailings at Porco were formerly dumped along the river, they are now discharged into a new tailings impoundment that complies with developed countries' standards — an impermeable layer of argillaceous and coarse material covering the dam prevents the contamination of groundwater (from acid mine drainage and the leaching of flotation reagents).

At the time of writing, Porco had not yet developed an environmental-protection program, which was to have included mitigation of pollution from its old tailings dam; environmental assessment of its current operations; and decommissioning and reclamation programs.

The striking expansion in production capacity at Inti Raymi, 3 500% between 1985 and 1993, reflects the development of an oxide heap-leaching project from 1985 to 1992 and the commissioning of a sulfide agitation-leaching project in 1993. Despite Inti Raymi's significant expansion of its heap-leaching project, it did not change its waste-disposal methods because the area has abundant land and low population density and the toxicity of the tailings is relatively low. Further expansions were rejected after an evaluation of ore reserves showed that additional economically exploitable, oxidized reserves were unavailable. Instead, Inti Raymi set out to find a partner (Battle Mountain Gold Company) to provide the capital and technology needed to develop and exploit its sulfide deposit. To exploit the sulfide deposit, Inti Raymi needed to introduce (for the first time in Bolivia) agitation leaching and enlarge its operation to 14 500 t/d (Ugalde 1992). The agitation-leaching project was expected to generate about 120×10^6 t of waste, 13 times the amount produced by the heap-leaching project. This sulfidic waste has a greater potential for producing acid mine drainage than the oxidized material does, and its management therefore merits special consideration. Inti Raymi prepared a plan for decommissioning, waste management, and reclamation, including reprocessing of the heap-leaching tailings (see Pero et al. 1992).

Two conclusions emerge from this analysis. First, significant improvements in firms' competitiveness — like those experienced by COMSUR and Inti Raymi in the 1980s — are likely to lead to these firms' forming partnerships with companies that have incorporated more efficient environmental practices into their mining operations. Thus, along its evolutionary path, a firm may be positively influenced in its environmental management by having a partnership with a more competitive company (such as COMSUR's partnership with RTZ). Second, the relationship between changes in pollution per unit of output and changes in production capacity resembles a discontinuous function. Incremental changes in a firm's production capacity have, at best, less significant effects on its pollution per unit of output. For example, Inti Raymi's water-management and waste-disposal methods did not change significantly as a result of the expansion of its heap-leaching project; and the same applies to COMSUR's introduction of flotation at Porco. However, a major change in a firm's production capacity (usually including a radical change in the production process) can significantly affect the firm's use of the environment (for example, Inti Raymi's agitation-leaching project). Reductions in pollution per unit of output in the mining industry may result not only from incremental technical change (as considered in the introductory sections) but also from radical technical change.

Lessons from contradictory evidence

The cases of CECOCA's Rosario California cooperative (1978–86) and COMIBOL's Colquiri operation (1977–92) provide contradictory evidence that fails to support the model. Although CECOCA's competitiveness increased between 1978 and 1986, this did not affect the pollution per unit of output of its Rosario California cooperative. This was principally because Rosario California had accumulated negligible technological capability and so achieved no increases in gold-recovery rates. Rosario California's failure to accumulate technological capability was due to its corporate strategy. The cooperative's main objective — to guarantee that all of its members work and partake of its profits equally — discouraged labour specialization and, in particular, the accumulation of knowledge and technical expertise. The cooperative developed only those work processes that its unskilled members could carry out. Moreover, as Rosario California's corporate strategy was formulated by the members' assembly, the cooperative suffered from a lack of leadership and was inefficiently managed. This type of development was reinforced by the policies of the technical and financial institutions that supported mining cooperatives' activities, such as the Mining Bank. These institutions granted loans according to appraisals that only considered investment in production capacity and had as their main concern a project's cash flow. The Mining Bank's technical assistance to the cooperatives was limited to corroborating the availability of reserves and a project's economic feasibility. Evidence from CECOCA indicates that a firm's reduction in pollution per unit of output requires not only an increase in its competitiveness but also minimal organizational conditions for the firm to accumulate management and technological capabilities.

At COMIBOL's Colquiri operation, pollution per unit of output has shown some rather paradoxical trends. Although COMIBOL's competitiveness declined, reagent-consumption and tin-recovery rates at Colquiri decreased or remained constant in 1990–92. The average water consumption per tonne treated was 30% less in 1989–92 than it had been in 1977–85. However, water recycling in the dressing process decreased only slightly in 1989–92 from that of 1977–88. Furthermore, Colquiri disposed of 13% less tailings in its tailings impoundment, thereby increasing the discharge of waste per tonne treated into natural streams. These contradictory trends were due to the advantageous environmental effects of incremental technical change and the negative effects of COMIBOL's reduction in production capacity, a result of its decline in competitiveness.

Although the reduction in production capacity at Colquiri (from 2 200 t/d to 1 000 t/d) was not as severe as that at Catavi or Unificada, it was nonetheless severe enough to make the need for water recycling less pressing than it had been in 1978–85. COMIBOL's environmental performance at Colquiri improved because of the redesign of Colquiri's concentrator between 1988 and 1989. This new mineral-processing circuit enabled the firm to reduce its energy and water consumption per tonne treated. The fact that technical change occurred despite the firm's decline in competitiveness and despite severe reductions in its investment in technological capability does not mean we have to reject the model, because technical change and reductions in pollution per unit of output may result from a firm's previous technological accumulation as well as its current investments in technological capability.

Summary of findings

The comparative analysis of the case study showed that the model's main prediction — that pollution per unit of output is inversely related to a mining firm's competitiveness — was not rejected. Empirical evidence confirmed that variations in the stocks of technological capability and production capacity are the primary transmission mechanisms between changes in competitiveness and changes in environmental performance. The case studies also showed, however, that these relationships involve more factors than anticipated by the model — in particular, the flow of foreign technology, management, and capital into the domestic economy and social factors such as unemployment are important. Accordingly, as long as actual competition resembles dynamic competition, a mining firm's efficiency in exploiting ore deposits significantly influences the extent to which it externalizes environmental costs.

Environmental costs and competitive efficiency

The principal result of this study is that a mining firm's ability to compete significantly affects its internalization of environmental costs. A mining firm's trajectory of competitiveness, therefore, conditions its trajectory of pollution per unit of output over time. Other things being equal, an improvement in a mining firm's competitiveness tends to reduce its pollution per unit of output, and a decline in its competitiveness tends to increase its pollution per unit of output. Thus, excessive environmental degradation originates not only from the inability of the market to adequately price environmental resources but also from a mining firm's competitive inefficiency. Moreover, the analysis showed that, under conditions of dynamic competition, the problem of the depletion of natural resources (user cost) is essentially related to the problem of environmental pollution (pollution costs). Figure 9 illustrates these findings. According to the conventional economics of static competition, the externalization of environmental costs is caused by the market's failure to adequately price environmental resources. In this study, by contrast, two variables determine a particular level of externalization of environmental costs: market failure and a mining firm's competitive efficiency.

Market failure prompts excessive environmental degradation in at least two ways. First, as identified in conventional economic theory, the market's failure to adequately price environmental resources may induce a firm to externalize some of its environmental costs. Second, the market's failure to provide insurance against the risks involved in the development of innovations and appropriation of technological assets embodied in human capital makes a firm underinvest in technological capability. Because technological accumulation is crucial to a firm's competitive efficiency and to internalizing environmental costs, such underinvestment in technological capability may aggravate environmental degradation. A firm's ability to compete also determines the degree to which it internalizes its environmental costs. This is because improvements in competitiveness encourage investment in technological capability and production capacity. A firm's lack of accumulated technological capability weakens its ability to innovate, which increases waste and losses of harmful substances per unit of output. Moreover, improvements in competitiveness and technical change prompt increases in a firm's production capacity, which encourage improvements in its management of the natural environment as the scarcity effects related to the use of environmental resources become more acute. Increases in a firm's production capacity will therefore lead to improvements in waste-disposal methods and reductions in the use and pollution of complementary environmental resources.

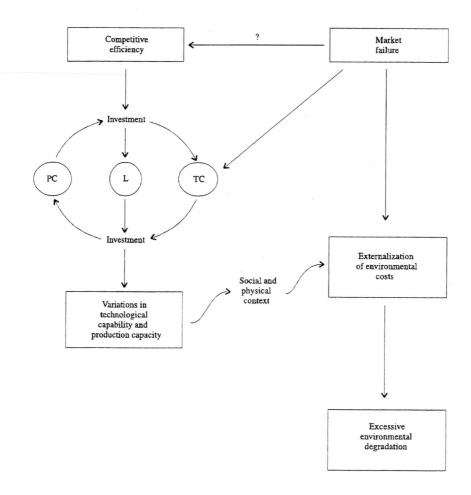

Figure 9. Factors underlying excessive environmental degradation. Note: ?, indeterminate; PC, production capacity; L, leadership; TC, technological capability.

The negative effect that a mining firm's inability to innovate has on its environmental performance is well illustrated by COMIBOL's failure to introduce cassiterite flotation and volatilization to recover tin. In the first stage, COMIBOL was unable to arrest the decline in its tin-recovery rates, and it lost the opportunities for accumulating production capacity and diversifying production that had been opened by the rising prices for tin and other metals in the 1970s. In the second stage, during the first half the 1980s, COMIBOL's decline in competitiveness (aggravated by the reduction in tin prices) led to a severe shrinkage in its production capacity. Catavi and Unificada, two of the most important of COMIBOL's tin operations, scaled down their operations dramatically and decreased their consumption of water. With water recycling no longer a necessity and the nearby rivers' capacities for receiving mining tailings no longer at risk of being overloaded, both operations discharged tailings directly into the river. Moreover, the severe reduction in COMIBOL's production capacity caused unemployment and thus encouraged the emergence of artisan mining operations. A significant increase in pollution per unit of output was the result — these new operations discharged tailings and toxic substances, such as sulfuric acid and fuels, into natural streams.

Conversely, evidence from Inti Raymi clearly illustrates how a mining firm's internalization of environmental costs may follow improvements in its competitiveness. In 1982–85, Inti Raymi started using a 400-t/d heap-leaching operation to exploit a massive gold and silver deposit. At that time, the firm had no reclamation plan at all. When heaps were exhausted, the firm neither thoroughly rinsed out the cyanide nor reclaimed the heaps. In 1986, Inti Raymi established an R&D department, which introduced several incremental technical changes. As a result, the firm optimized the size of the mineral grain, improved the watering of heaps, and greatly enhanced the cycle of the cyanide solution, including thoroughly retrieving the solution from the old heaps. Overall, these changes positively affected Inti Raymi's environmental performance. They significantly increased gold-recovery rates and reduced cyanide consumption per tonne treated by 70%. The remarkable improvement in Inti Raymi's competitiveness led it to enlarge its production capacity from 4 000 t/d to 14 500 t/d. Inti Raymi increased its reserves by 1 200%, and it developed a new mining project (agitation leaching). This project was a watershed in the Bolivian mining industry because, for the first time, a project had included, from the project-feasibility stage, the design for a comprehensive mitigation and reclamation program.

Investment efficiency

The case study indicates that the model is limited by its failure to distinguish between investment and investment efficiency. It fails to consider the possibility that a particular amount of money invested in production capacity and the same amount invested in technological capability might produce quite different outcomes. The variable L (leadership; referring to a firm's ability to manage its investments) is introduced in Figure 9 to handle this distinction. A mining company's ability to manage its investments is essential to its competitive efficiency. This is particularly true of investments in technical change, as successful technical change involves management of complex processes over time. The case study showed that a firm's capabilities to adapt to technical change to continually improve its technologies at the operation site were deciding factors in the firm's competitive success. For instance, the change in COMSUR's Porco plant design, from lead–zinc differential float to bulk flotation, was vital to reaping the advantages of its new concentrator. Moreover, the remarkable improvements in the heap-leaching project at Inti Raymi were possible because of further investments in technological capability to improve the production parameters at the mining site and to deal with bottlenecks and problems. In contrast, COMIBOL and CECOCA were unable to reap the potential benefits of the technical changes they introduced into their production processes. COMIBOL mismanaged tin volatilization and cassiterite flotation, and CECOCA lost significant amounts of fine gold for lack of technical changes to improve gold-recovery rates. Thus, both firms failed to continually improve the production process at the operation site, and this was due to a lack of knowledge, expertise, and appropriate institutional linkages.

It seems that to manage technical change, a mining firm requires efficient management and coordination of all its activities at the corporate level. In the heap-leaching project at Inti Raymi, for example, the efficient coordination of on-going production activities and investment projects was crucial to accruing the advantages of the incremental technical changes introduced by its R&D department. Moreover, efficient management was a key factor in the successful completion of the agitation-leaching project. In contrast, the failure of the cassiterite-flotation process at COMIBOL's Catavi operation was due to the management's lack of leadership and commitment to organizational change, such as redeploying personnel, reducing the work force, training human resources, finding financial sources, and making long-term plans to transform Catavi from an underground to a surface operation. So, although the cassiterite-flotation process was technologically accomplished, it had minor economic benefits for COMIBOL. At Unificada, the cause of failure was not a lack of the technical expertise and knowledge needed to manage the tin-volatilization process but a lack of the expertise and knowledge required to negotiate with technology suppliers; arrange the tasks of a complex investment plan according to a schedule; cope with pitfalls and unforeseen circumstances; and coordinate the supply of low-grade tin concentrates. In short, the failures of the tin-volatilization process at Unificada and the cassiterite-flotation process at Catavi were largely due to management shortcomings.

As COMIBOL and CECOCA illustrate, the impact that investments in technical change have on a firm's ability to compete depends on leadership, knowledge, expertise, and institutional linkages related to the management of not only technical processes, such as product design, process engineering, and R&D, but also organizational procedures, such as marketing, finance, planning, and human-resource development.

Implications for policy and further research

Although this study is based on the Bolivian mining industry, it has policy implications of a much wider scope, as the causal relationships and factors identified
are likely to be relevant in other industries and in other countries. An important
policy implication of this research is that a nation needs to integrate environmental, technological, and industrial policies if its economy is to follow an environmentally sensitive pattern of development (which reinforces the conclusions of
Warhurst [1993, 1994] in reviewing environmental regulations in the mining
industry). Understanding the factors underlying companies', industries', and the
economy's abilities to compete is of great importance in designing policies that
in the long run promote the rational use of the natural environment as well as
economic growth. Thus, a study of a mining firm's ability to compete is needed
to develop a long-term environmental policy for the mining industry, which may
be the main policy implication of this study.

Reconsidering the model

To improve the model, one must consider at least two issues. First, the model
deals with changes in competitiveness as an exogenous variable, so its predictive
value (and potential usefulness for policy design) is limited: investments in
production capacity and technological capability affect not only a firm's pollution
per unit of output but also its competitiveness. This limitation means that the evolutionary features of dynamic competition are defined *a priori*, rather than arising
from the dynamic effects of competition itself. If this variable is made internal to
the system of equations, one will be able to systematically consider the influences
of science, technology, and market contexts on competitiveness and environmental
performance. This might provide a tool for designing long-term environmental and
mineral policies and open a different perspective on the analysis of economic
processes.

Second, the model leaves uncertain the effect of radical technical change
on the relationship between a mining firm's changes in competitiveness and
changes in its pollution per unit of output. Because radical technical change may
have a significant influence on the development of the mining industry in the long
run, a historical study of the technological development of the mining industry is
advisable. This work should aim to assess the long-term effects of radical technical change on competitiveness and pollution per unit of output, rather than its
immediate (short-term) effects. This will also be relevant to policy design —
policies can be tailored to promote both the innovations that are potentially the

most beneficial to the environment and the competitive and regulatory conditions that will allow these technologies to fulfil their potential.

Competitive efficiency and the pollution-prevention principle

The inability of market forces to properly price environmental resources has been the main justification for government intervention through environmental policy. Environmental policy has, therefore, mainly regulated the disposal of industrial residues and wastes. Under the static approach to competition, the trade-off between environmental protection and the promotion of industry has been considered a major constraint on pollution reduction. The recent shift toward the pollution-prevention principle has been accompanied by a shift from policies designed to ameliorate pollution toward those based on a preventive agenda. The intention is for industry to reduce or eliminate waste at source, rather than merely treating and disposing of it (see, in the context of the minerals sector, Anderson and Purcell [1994]).

Although the pollution-prevention approach is an advance in dealing with environmental degradation, there are two constraints on the development of a still more efficient environmental policy. First, the analysts tend to consider environmental policy a social mechanism for internalizing externalities, as it is thought that environmental degradation originates in the market's inability to appropriately price environmental resources. Second, the policymakers have not considered the need for technology policies to complement environmental policies, as it is assumed that competition is static. Accordingly, environmental policy has neglected the relationship between a company's ability to innovate and its environmental performance (Warhurst 1993).

This study provides empirical evidence for the relationship between a company's ability to innovate and its environmental performance. Further, it shows that at least for the mining industry, excessive environmental degradation is also caused by competitive inefficiency, which so far the pollution-prevention approach has not considered. Policy mechanisms should be developed to prevent pollution arising from the decline in competitiveness of industrial activities (such as COMIBOL's Catavi operation). Although this study suggests that progress toward the optimal use of the environment is compatible with industrial development, this does not imply a blind confidence in spontaneous or automatic mechanisms of any kind. Industrial evolution is a social rather than natural process, so the challenge will be to establish institutions, practices, and policies to make such progress feasible. In this regard, this study has identified promising new avenues to explore and examine, rather than providing solutions or answers.

APPENDIX 1

CONTRIBUTORS

Liliana Acero
CIS Director
Argentina
and
Associate Professor
Department of Sociology
University of Massachusetts
and Brown University
Providence, RI, USA

Teresinha Andrade
Centro de Tecnologia Mineral
Rio de Janeiro, Brazil

Isabel Castañeda-Hurtado
Instituto de Salud Popular
Pontificia Universidad Católica del Perú
Peru

Juanita Gana
Centro de Estudios del Cobre y de la Mineria
Chile

Maria Hanai
Technical and Scientific Development Analyst
Ministry of Science and Technology
Brazil
and
Researcher
NAMA/FEA
University of São Paulo
São Paulo, Brazil

Gustavo Lagos
Faculty of Engineering
Catholic University of Chile
Vicuña Mackenna 4860, Santiago, Chile

Ismael Fernando Loayza
Ministerio de Energia y Minas
La Paz, Bolivia

Alfredo Núñez-Barriga
Instituto de Salud Popular
Pontificia Universidad Católica del Perú
Peru

Patricio Velasco
Centro de Estudios del Cobre y de la Mineria
Luis Thayer Ojeda 059, Depto. 43, Santiago, Chile

Alyson Warhurst
Professor and Director
Mining and Environment Research Network
School of Management
University of Bath
Bath, UK BA2 7AY

ACRONYMS AND ABBREVIATIONS

ABAL	Associação Brasieira da Indústrio do Alumínio (Brazilian association of aluminum producers)
ALBRAS	Alumínio Brasileiro S.A. [Brazil]
ALCOMINAS	Compañía Mineira de Alumínio [Brazil]
ALUNORTE	Alumina do Norte do Brasil [Brazil]
ANOVA	analysis of variance
ANR	National Assembly of University Vice-Chancellors [Peru]
asl	above sea level
BAT	Best Available Technology standards
BISA	Buenaventura Ingenieros S.A. [Peru]
CAA	*Clean Air Act* [United States]
CAF	Andean Finance Corporation
CDN	National Defence Council [Brazil]
CECOCA	Central Local de Cooperativas Cangallí [Bolivia]
CEQ	Council on Environmental Quality [United States]
CERCLA	*Comprehensive Environmental Response, Compensation, and Liability Act*
CESBRA	Compañía Estanifera do Brasileira [Brazil]
CETEC	Fundação Centro Tecnológico (central technological foundation) [Brazil]
CETEM	Centro de Tecnologia Mineral (centre for mineral technology) [Brazil]
CEZ	critical environmental zone
CEZM	CEZ with mining as main source of environmental degradation
CIGA	Centro de Investigacion en Geographia Aplicada (centre of applied geographic research) [PUCP]
CIMA	internal environmental commission [within each Brazilian mining companiy]
CODELCO	Corporación Nacional del Cobre (national copper corporation) [Chile]

COFIDE	Coorporación Financier de Desarrollo S.A. (development finance corporation)
COMIBOL	Corporación Minera de Bolivia [Bolivian mining corporation]
COMSUR	Compañía Minera del Sur [Bolivia]
CONAMA	Comisión Nacional del Medio Ambiente (national commission of the environment) [Brazil, Chile]
CONAPMAS	National Council for the Protection of the Environment [Peru]
COREMA	Comisión Regional del Medio Ambiente (regional commission of the environment) [Chile]
CVRD	Compañía Vale do Rio Doce [Brazil]
CWA	*Clean Water Act* [United States]
DEA	Division of Environmental Affairs [Centromín, Peru]
DFM	Direccion de Fiscalización Minera (directorate of mining control) [Peru]
DGA	Direccion General de Agua (general directorate of waters) [Chile]
DGAA	Direccion General de Asuntos Ambeintal (general directorate for environmental affairs) [Peru]
DIGESA	Direccion General de Salud Ambiental (general directorate for environmental health) [Peru]
DNPM	Departmento Nacional de Produção Mineral (national deparment of mineral production) [Brazil]
DSSE	Division of Stationary Source Enforcement [EPA]
EBESA	Empresa Brasileira de Estanho S.A. (Brazilian tin company)
ECOTEC	Ecología y Tecnología Ambiental [Peru]
EFL	environmental-framework legistation [Chile]
EIA	environmental-impact assessment
EIA	estudio de impacto ambiental (environmental-impact study)
EID	environmental-impact declaration
EIS	environmental-impact study
ENAMI	Empresa Nacional de Minería (national mining company) [Chile]
EPA	Environmental Protection Agency [United States]
FACA	*Federal Advisory Committee Act* [United States]
FEAM	Fundação Estadual do Meio Ambiente (state foundation for the environment) [Brazil]
FEEMA	Fundação Estudual de Engenharia do Meio Ambientestate (foundation for environmental engineering) [Brazil]
FIA	Faculty of Environmental Engineering [UNI]
FIMMG	Faculty of Mining, Metallurgy and Geological Engineering [UNI]

IBAMA	Instituto Brasileiro do Meio Ambiente e dos Recursos Naturais e Renováveis (Brazilian institute of the environment and renewable natural resources)
IBGE	Instituto Brasileiro de Geografia e Estatitica (Brazilian institute of geography and statistics)
IBRAM	Instituto Brasileiro de Mineração (Brazilian institute of mining)
ICME	International Council on Metals and the Environment
IDB	Inter-American Development Bank
IDRC	International Development Research Centre
IEA	Institute for Environmental Studies [PUCP]
IFC	International Finance Corporation [World Bank]
IIC	Inter-American Investment Corporation [IDB]
IIMM	Instituto de Investigaciones Minero Metalúrgico (institute for the study of mining and metallurgy) [Bolivia]
IMC	International Management Centres
INGEMMET	Instituto Geológico, Minero y Metalúrgico (geological, mining, and metallurgical institute) [Peru]
INRENA	Instituto Nacional de Recursos Naturales (national institute of natural resources) [Peru]
INSO	Instituto Nacional de Salud Ocupacional (national institute of occupational health) [Peru]
IRPJ	Imposto de Renda Pessoa Jurídica (rent tax for mines)
JICA	Japan International Cooperation Agency
LAGESA	Laboratorio Geotécnica S.A. [Peru]
LDC	least-developed country
LME	London Metal Exchange
MC	marginal cost of pollution
MEM	Ministerio de Energía y Minas (ministry of energy and mines) [Peru]
MERN	Mining and Environment Research Network
MRN	Mineraçao Rio do Norte [Brazil]
MSB	marginal social benefits
MSC	marginal social costs
NAAQS	National Ambient Air Quality Standards [EPA]
NEPA	*National Environmental Protection Act* [United States]
NIE	newly industrialized economy
NSO	Non-ferrous Smelter Order
NSPS	New Source Performance Standards [EPA]
OECD	Organisation for Economic Co-operation and Development

ONERN	Oficina Nacional de Evaluación de Recursos Naturales (national office for the evaluation of natural resources) [Peru]
OSW	Office of Solid Waste [EPA]
PAMA	Programa de Adecuación y Manejo Ambiental (environmental-management and adequation program) [Peru]
PDC	Policy Dialogue Committee [EPA]
PRDA	plan for recovering degraded area
PUCP	Pontificia Universidad Católica del Perú (Catholic University of Peru)
R&D	research and development
RCRA	*Resource Conservation and Recovery Act* [United States]
RIMA	relatório de impacto ambiental (environmental-impact report)
RTZ	Kennecott Corporation
SERNAGEOMIN	Servicio Nacional de Geología y Minería (national geological and mining service) [Chile]
SIMSA	San Ignacio de Morococha [Peru]
SIP	State Implementation Plan [United States]
SISNAMA	Systema Nacional de Meio Ambiente (national system for the environment) [Brazil]
SMCRA	*Surface Mining Control and Reclamation Act* [United Steates]
SNIEE	Sindicato Nacional da Indústria de Extração do Estanho (national union of the tin-mining industry)
SPCC	Southern Peru Copper Company
SX–EW	solvent extraction – electrowinning
TQC	total quality control
UNCED	United Nations Conference on Environment and Development
UNI	Universidad Nacional de Ingeniería (national university of engineering) [Peru]
USBM	United States Bureau of Mines
USD	United States dollar(s)
WGA	Western Governors Association [United States]

BIBLIOGRAPHY

ABAL. 1990. Statistics yearbook. ABAL. Comissao do Meio Ambiente, Sao Paulo, Brazil.

——— 1991. A legislaçao Ambiental e suas interfaces com a industria do aluminio no Brasil. Comissao do Meio Ambiente, São Paulo, Brazil.

Acero, L. 1993a. Environmental management in the bauxite, alumina and aluminium industry in Brazil. Warhurst, A., ed., Science Policy Research Unit, The University of Sussex, Brighton, UK.

——— 1993b. The case of bauxite, alumina and aluminium in Brazil. Mining and Environment Research Network, Bath, UK. Working Paper Series, No. 1, Feb. 99 pp.

ACS (Asociación Chilena de Seguridad). 1991. Anales de la Primera Jornada sobre Arsenicismo laboral y ambiental. II Región, Aug.

Acton, J.P.; Dixon, L.S.; et al. 1992. Superfund and transaction costs: the experiences of insurers and very large industrial firms. RAND Institute for Civil Justice, Santa Monica, CA, USA.

Alumar. 1990–92. Financial records. Alumar, Maranho, Brazil.

——— 1991. Administraçao industrial. Alumar, Maranho, Brazil.

Andaluz, A.; Valdez, W. 1987. Derecho Ecológico Peruano. Inventario Normativo. GREDES.

Anderson, F.R.; Mandelker, D.R.; Tarlock, A.D. 1984. Environmental protection: law and policy. Little, Brown, Boston, USA. 978 pp.

Anderson, K.; Purcell, S., ed. 1994. Proceedings of the International Conference on Pollution Prevention. Colorado School of Mines, Golden, CO, USA.

Anon. 1991. Project 88 — round II. Incentives for action: designing market-based environmental strategies. Washington, DC, USA.

—— 1993a. INCO to sell stake in TVX Gold Unit to Underwriters for $29,4 million. The Wall Steet Journal, New York, NY, USA, 29 Jun.

—— 1993. Gestión, Lima, Peru, 11 Nov, p. 28.

—— 1993. Gazeta Mercantil, 21 Jun, p. 19.

—— 1994. Gestión, Lima, Peru, 3 Feb, p. 32.

Araújo Neto, H. 1991. Relatório Suscinto do Projeto Ouro–Gemas 1990–1991 [succint report on the gold–gems project: 1990–1991]. CPRM–DNPM, Brasília, Brazil.

Ashford, N.A. 1991. Legislative approaches for encouraging clean technology. Technology and Industrial Health, 7(516), 335–345.

Ashford, N.A.; Heaton, G.R. 1979. The effects of health and environmental regulation on technological change in the chemical industry. In Hill, C.T., ed., Government regulation and chemical innovation. American Chemical Society, Washington, DC, USA.

Auty, R.; Warhurst, A. 1993. Sustainable development in mineral exporting economies. Resources Policy, Mar.

Barbier, E. 1989. Economics, natural resource scarcity, and development: conventional and alternative views. Earthscan Publications Ltd, London, UK.

—— 1991. Environmental degradation in the Third World. In Pearce, D.; Barbier, E.; Markandya, A., ed., Blueprint 2: greening the world economy. Earthscan Publications, London, UK.

Barnett, A. 1993. Technical co-operation, technology transfer and environmentally sustainable development: a background paper for the DAC Working Party on Development Assistance and the Environment. Organization for Economic Co-operation and Development, Paris, France.

Barreto, L.; Coelho Neto, J.S. 1993. O arcaboço jurídico da mineração. CETEM–CNPQ, Rio de Janeiro, Brazil. Relatório de Projeto.

Bell, R.M. 1990. Continuing industrialisation, climate change and international technology transfer, a report prepared in collaboration with the Resource Policy Group, Oslo, Norway. Science Policy Research Unit, University of Sussex, Brightton, Sussex, UK.

Berbet, C.O. 1988. Gold. In Departmento Nacional de Produção Mineral (department of mineral production), ed., Principais Depósitos Minerais do Brasil [main mineral deposits of Brazil]. Brasília, Brazil. pp. 290–411.

Bórquez, J.M. 1993. Introducción al derecho ambiental chileno comparado, Editorial Jurídica, Santiago, Chile. 144 pp.

Brasil Mineral. 1985. Raposos opera em Janeiro [Raposos starts its operation in January]. Brazil Mineral (21), 12–16.

————. 1989. Pionerismo em preservação ambiental [Pionerism with environmental preservation]. Brazil Mineral 7(72), 64–66.

Brewis, T. 1991. Flotation cells. Mining Magazine (Jun 1991), 383–393.

Burckle, J.; Worrel, C., III. 1981. Comparison of environmental aspects of selected nonferrous metals production technologies. *In* Chatwin, T.D.; Kikumoto, N., ed., Sulfur dioxide control in pyrometallurgy: proceedings of a symposium. 110th AIME Annual Meeting, Chicago, IL.

Canelas, A. 1981. Quiebra de la Mineria Boliviana? Editorial Los Amigos del Libro, La Paz; Cochabamba, Bolivia.

Centromín. 1992. Programa de Adecuación y Manejo Ambiental: Proyectos técnicos para la reducción de la problemática ambiental. Vol. I: Resumen Ejecutivo; Vol. II: Fundición y Refinerías, La Oroya; Vol. III: Unidades de Producción Minera.

———— n.d. Memorias [annual reports]. Various years.

CEQ (Council on Environmental Quality). 1993. 23 annual report of the Council on Environmental Quality. CEQ.

Cerqueira, F. 1992. Formação de recursos humanos para a gestão ambiental. Revista Administração Pública, 1(26), 50–55.

Charles River Associates Inc. 1971. The economic effects of pollution controls on the nonferrous metals industry: copper. Prepared for the Council on Environmental Quality. Washington, DC, USA.

Chiang, J.; Cornejo, P.; Lopez, J.; Romano, S.; Pascual, J.; Cea, M. 1985. Determinación de cadmio, cobre manganeso, plomo, hierro, cinc, y arsénico en sedimento atmosférico en la zona de Quintero, V región, Valparaíso. Bol. Soc. Chil. de Química, 30(3), 139–158.

Cole, H.S.D.; Freeman, C.; Jahoda, M.; Pavitt, K.L.R., ed. 1973. Thinking about the future: a critique of limits to growth. Chatto & Windus, London, UK.

CONAMA (Comisión Nacional del Medio Ambiente). 1992. National Commision for the Environment, Legislative database.

CONAPMAS (National Council for the Protection of the Environment). 1988. Informe sobre Registro de Denuncias por Contaminación Ambiental en el Sector Minero. Alcántara, M.

Cook, M.A.L. 1989. Copper smelting and refining. The cold wind of competition begins to blow. Natural Resources Forum, 13(2).

Coppel, N. 1992. World-wide minerals and metals investment and the environment, 1980–1992. Unpublished report for RTZ Corporation Plc.

Cota, R.G.; et al. 1986. Relations between the "Garimpo" and the Land Property Structure: the example of Marabá. Pará Desenvolvimento, Belém (119).

Couto, R.C.S.; Câmara, V.M.; Sabroza, P.C. 1989. Mercury intoxication: preliminary results in two "garimpo" areas — PA. Cadernos de Saúde Pública, 4(3), 301–315.

CPD (Concertación por la Democracia). 1989. Electoral program. CPD, Chile.

Crandall, R. 1981 The US steel industry in recurrent crisis. The Brookings Institute, Washington, DC, USA.

————— 1983. Controlling industrial pollution. The Brookings Institute, Washington, DC, USA.

CRS. 1981. A congressional handbook on US materials import dependency/vulnerability. Report prepared for the Subcommittee on Economic Stabilization of the Committee on Banking, Finance and Urban Affairs, House of Representatives. CRS, Washington, DC, USA.

————— 1984a. Decline in competitiveness of the US copper industry. Report prepared for the use of the Subcommittee on Oversight and Investigations of the Committee on Energy and Commerce, US House of Representatives. CRS, Washington, DC, USA.

————— 1984b. The status of environmental economics: an update. Prepared by the Environmental Policy Division for the Committee on Environment and Public Works, US Senate. CRS, Washington DC, USA.

Dames and Moore Consultants. 1991. Calidad de agua en hoyas hidrográficas de la III, IV, V, VI regiones y Región Metropolitana. Dames and Moore Consultants, Santiago, Chile.

de la Fuente, J. 1993. Politica Tecnologica de COMIBOL. Centro de Estudios Mineria y Desarrollo, La Paz, Bolivia. Unpublished working paper.

DEM–DNPM. (Mineral Economics Department–National Department of Mineral Prodution of the Ministry of Mines and Energy of Brazil). n.d.

Dower, R. 1991. Hazardous waste policy. *In* Portney, P., ed., Public policies for environmental protection. Resources for the Future, Washington, DC, USA.

DIGESA (Direccion General de Salud Ambiental). n.d. Estudio de contaminación y preservación del Río Rímac. DIGESA, Lima, Peru.

DMMH (Division of Malaria of the Ministry of Health). 1991. Correio Brasiliense, Brasília, Brazil, p. 10.

DNPM (Departmento Nacional de Produção Mineral). 1990. Anuário Mineral Brasileiro. 1980–1990. DNPM, Brasília, Brazil.

———— 1992. Avaliação da carga tributária sobre o setor mineral. DNPM, Brasília, Brazil. Estudos de Política e Economia Mineral No. 6.

———— n.d. Sumáio Mineral. DNPM, Brasília, Brazil. Various years.

Doeleman, J. 1991. Environment and development. Conceptual and normative observations. Development and International Co-operation, 8(13), 13–41.

EPA (Environmental Protection Agency). 1990a. Strawman II. Recommendations for a regulatory program for mining waste under Subtitle D of the *Resource Conservation and Recovery Act*. EPA, Office of Solid Waste, 21 May.

———— 1990b. Report of the Environmental Protection Agency Economic Incentives Task Force.

———— 1991. Report and recommendations of the technology innovation and economics committee. Permitting and compliance policy: barriers to US environmental technology innovation. National Advisory Council for Environmental Policy and Technology. EPA 101/N-91/001.

Epstein, A.L. 1992. Estratégias empresariais e o novo paradigma ambiental. CETEM–CNPQ, Rio de Janeiro, Brazil. Relatório de Consultoria.

Freeman, C. 1992. Values, economic growth and the environment. *In* Freeman, C., ed., The economics of hope: essays on technical change, economic growth and the environment. Pinter Publishers, London, UK; New York, NY, USA.

Freeman, C.; Jahoda, M. ed. 1978. World futures. Martin Robertson, London, UK.

Galaz J. 1993. Revista Minerales, 48(201).

GAO (General Accounting Office). 1986. Sulphur dioxide emissions from non-ferrous smelters have been reduced. Report to the Chairman, Subcommittee on Oversight and Investigations, Committee on Energy and Commerce, House of Representatives. GAO/RCED-86-91.

Garret, J. 1968. La Revolucion Industrial de la Mineria. *In* Primer Simposio Internacional de Concentracion del Estano. Universidad Tecnica de Oruro, Oruro, Bolivia.

Garrido Filha, I. 1983. Garimpos de Cassiterita. Pesquisa geográfica em Goiás. University of São Paulo, Brazil. PhD dissertation.

—— 1992. A Mineração de Cassiterita no Brasil. CETEM–CNPQ, Rio de Janeiro, Brazil. Relatório de Consultoria.

Gestión. 1993. (Lima newspaper.) 11 November, 28.

Goerold, W.T. 1987. Environmental and petroleum resource conflicts. A simulation model to determine the benefits of petroleum reduction in the Arctic National Wildlife Refuge, Alaska. Materials and Society, 11(3).

Gomez, M.; Duffy, R.; Trivelli, V. 1979. At work in copper. Occupational health and safety in copper smelting. Inform.

Gonzalez, S. 1992. Riesgo Ambiental para los suelos de Chile, Instituto de Investigación Agropecuaria. Report La Platina 70.

Gonzalez, S.; Bergqvist, E. 1985. Evidencias de contaminación con metales pesados en Puchuncaví. 4th Symposium on Pollution, Instituto de Investigación Agropecuaria (INIA, national institute for agricultural and cattlery research, Aug. INIA.

Graves, A.P. 1991. Globalisation of the automobile industry: the challenge for Europe. *In* Freeman, C.; Sharp, M.; Walker, W., ed., Technology and the future of Europe. Pinter Publishers, London, UK. pp. 261–282.

Gulley, D.; Macy, B. 1985. Benefits and costs of environmental compliance: a survey. *In* Vogely, W., ed., Economics of the minerals industries. 4th ed. American Institute of Mining, Metallurgical, and Petroleum Engineers, New York, NY, USA.

Habicht, F.H., II. 1992. EPA definition of pollution prevention. Environmental Protection Agency, Washington, DC, USA. Internal memo from the Deputy Administrator to all employees, 28 May.

Hacon, S. 1990. Mercury contamination in Brazil, with emphasis on human exposure to mercury in the Amazon Region. A Technical Report. Financiadora de Estudos e Projectos, Rio de Janeiro, Brazil. 69 pp.

Hahn, R. 1989. How the patient followed the doctor's orders. Journal of Economic Perspectives, (6).

Hahn, R.; Stavins, R. 1991. Incentive-based environmental regulation: a new era from an old idea? Ecology Law Quarterly, 18(1).

Haliechuck, R. 1993. Inco sells its TVX Gold Inc. stake. The Toronto Star, 29 Jun.

Hall, S. 1991. Developing flotation technologies. Mining Magazine (Jun), 379–381.

Hanai, M. 1993. Industrial mining, "garimpos" gold prospectors and the environment in Brazil. Mining and Environment Research Network. Progress Report, Jul.

Hotelling, H. 1931. The economics of exhaustible resources. Journal of Political Economy, 39, 137–175.

Housman, V.E. 1990. EPA's tailored regulations under the *Resource Conservation and Recovery Act* for the management of mining wastes. US Evironmental Protection Agency, Washington, DC, USA.

Housman, V.E.; Walline, R. 1990. EPA's waste regulations. Mining Journal, 24 Aug.

Huayhua, J.C. 1993. Technological progress in metallurgical complexes and adequation to environmental control: possibilities for its application in Centromín Perú. Paper presented at the Conversatorio de Abhorro y Sustitución de Energiá en Operacione Minero–Metalúrgicas, 12 May 1993, Centromín, La Oroya, Peru.

Hurrell, A.; Kingsbury, B., ed. 1987. The international politics of the environment. Clarendon Press, Oxford, UK.

IBGE (Instituto Brasileiro de Geografia e Estadistica). n.d. Pesquisa industrial mensal: emprego, salário e valor da produção. IBGE, Brazil.

IIMM (Instituto de Investigaciones Minero Metalurgicas). 1977/78. Water pollution control research proposal. IIMM, Oruro, Bolivia. Unpublished report.

INEI (Instituto Nacional de Estadística e Informática). 1993. Perú: Compendio Estadístico 1992–93. 3 Vols. INEI, Lima, Peru.

IRM (Intendencia de la Región Metropolitana.) 1988. Estudio del río Mapocho. IRM, Cile.

Irwin, F.H. 1989. Could there be a better law? EPA Journal, 15(4).

Jacobs, M. 1991. The Green economy. Pluto Press, Concord, USA.

Jordan, R.; Warhurst, A. 1992. The Bolivian mining crisis. Resources Policy, 18(1): 9–20.

Junac (Junta del Acuerdo de Cartagena). n.d. Catálogo de Frabricantes Andinos de Maquinaria y Equipo Minero.

Kemp, R.; Soete, L. 1990. Inside the "green box": on the economics of technological change and the environment. In Freeman, C.; Soete, L., ed., New explorations in the economics of technical change.

Kennecott Corporation. 1992. Press release.

Kimball, D.; Moellenberg, D. 1990. An industry view of the regulation of mining wastes and other solid wastes under Subtitle D of the Resource Conservation and Recovery Act. Paper presented at the SME Annual Meeting, 26 Feb–1 Mar, Salt Lake City, UT.

Kisic, D. 1993. Asesoría Financiera nacional en privatización de Hierro Perú. Revista Minería 220, 221 (Nov 1992–Feb 1993).

Kneese; Schultze. 1978. Pollution, prices and public policy. The Brookings Institute, Washington, DC, USA.

Kopp, R.J.; Smith, V.K. 1989. Benefit estimation goes to court: the case of natural resource damage assessments. Journal of Policy Analysis and Management, 8(4), 593–612.

Lacerda, D.; et al. 1987. Mercury contamination in Amazônia: preliminary evaluation of the Madeira River. Annals of the 1st Brazilian Congress of Geochemistry, Rio de Janeiro, Brazil, pp. 165–169.

Lagos, G.E. 1989. Preservación de un equilibrio ambiental en la explotación de los recursos no renovables. Proceedings of the 3rd Chilean Scientific Environmental Meeting, Sep, Centro de Investigación y Planificación del Medio Ambiente.

—————— 1990. Análisis de la situación del medio ambiente en relación con la minería Chilena. Centro de Investigación y Planificación del Medio Ambiente.

—————— 1992. Mining and environment: the Chilean case. Mining and Environment Research Network. Working Paper Series, Mar.

———— 1993. Study of the DGA database for copper content of all rivers in Chile from the First to the Sixth Region during the period 1985–1990. Comisión Chilena del Cobre.

Lagos, G.E.; Noder, C.; Solari, J. 1991. La situación Jurídica Institucional en el Area Minería y Medio Ambiente. Ministry of Mining, Santiago, Chile.

Loayza, F. 1993. Environmental management of mining companies in Bolivia: implications for environmental and industrial policies aiming at sustainable growth in low-income countries. Mining and Environment Research Network. Progress Report, Jul.

Larsen, J. 1981. Comparative philosophy of environmental regulations. *In* Chatwin, T.D.; Kikumoto, N. ed., Sulfur dioxide control in pyrometallurgy: proceedings of a symposium. 110th AIME Annual Meeting, Chicago, IL.

Leonard, J. 1988 Pollution and the struggle for the world product. Multinational corporations, environment and international comparative advantage. The Conservation Foundation, Cambridge University Press, Cambridge, UK.

Liroff, R.A. 1989. The evolution of transferable emission privileges in the US. The Conservation Foundation. Paper prepared for the workshop Economic Mechanisms for Environmental Protection, 26–29 Sep, Poland.

Luna, R.; Lagos, G.E. 1990. Evaluación del estado de las aguas del río Aconcagua contaminadas por acción de origen minero. Report for Centro de Investigación y Planificación del Medio Ambiente.

———— 1992. Metodología para el establecimiento de normas para efluentes metalúrgicos en la hoya hidrográfica de Copiapó. Proceedings of the 4th Chilean Environmental Scientific Meeting, Aug, Centro de Investigación y Planificación del Medio Ambiente.

MacDonnell, L. 1988. Regulation of wastes from the metals mining industry: the shape of the things to come. Mineral Processing and Extractive Metallurgy Review, 3.

———— 1989. Government mandated costs: The regulatory burden of environmental, health and safety standards. Resources Policy (Mar).

Mackenzie, B.; Dogget, M. 1991. Potencial Econômico da Prospecção e Pesquisa de Ouro no Brasil [economic potential of gold prospection and exploration in Brazil]. Departmento Nacional de Produção Mineral, Brasília, Brazil. Série Estudos de Política e Economia Mineral 4. 162 pp.

Malm, O., Pfeiffer, W.C., Souza, C.M.; Reuther, R. 1990. Mercury pollution in the Madeira river Basin, Brazil. Ambio, 19, 11–15.

Maron, M.; Silva, A.R.B. 1984. Perfil Analítico do Ouro [gold analytical overview]. Departmento Nacional de Produção Mineral, Brasília, Brazil. Bulletin 57, p. 143.

Martinelli, I.A.; Ferreira, J.R.; Forsberg, B.R.; Victoria, R.L. 1988. Mercury contamination in the Amazon: a gold rush consequence. Ambio, 17, 252–254.

McGraw-Hill. 1982. Annual survey of pollution control expenditures. Economics Department, McGraw-Hill Publications Company.

Meadows, D.H.; Meadows, D.L.; Randers, J.; Behrens, W.W., III. 1972. The limits to growth, Pan, London, UK.

Mikesell, R. 1987 Non fuel minerals. Foreign dependence and national security. Ann Arbor, The University of Michigan Press, Ann Arbor, MI, USA.

Milliman, S.R.; Prince, R. 1989. Firm incentives to promote technological change in pollution control. Journal of Environmental Economics and Management, 17, 247–265.

Minería y Desarrollo. 1992. Bulletin. Centro de Estudios del Cobre y de la Mineria, Santiago, Chile.

────── 1993. Bulletin. Centro de Estudios del Cobre y de la Mineria, Santiago, Chile.

Minérios: extração e processamento. 1991. 110 km de poços executados [110 km of shafts constructed]. Brazil Mineral 15(169), 46–4.

Mining Magazine. 1991. Scuddles' innovative recruitment. Mining Magazine, 164(1), 7.

Moore, J. 1986. The changing environment. Springer-Verlag, London, UK.

MRN (Mineração Rio do Norte). 1988. The Environmental Master Plan from Port Trombetas. Vol. 1: Protection and recoavery of the environment. MNR, Rio de Janeiro, Brazil.

────── 1990. Annual report. MRN, Rio de Janeiro, Brazil.

Muñoz, L.; Lagos, G.E. 1990. Análisis cuantitativo de la calidad de las aguas de la cuenca del Elqui. Report for Centro de Investigación y Planificación del Medio Ambiente.

Nappi, C. 1990. Public policy and competitiveness: the case of the bauxite–alumina–aluminium industry. École des hautes études commerciales, Montréal, PQ, Canada.

Navin, T. 1978 Copper mining and management. University of Arizona Press, Tucson, AZ, USA.

Núñez. 1991. Technical Change in Mineral Processing and Extractive Metallurgy, and its Implications for Environmental Control in Developing Countries: the cases of Copper, Lead and Zinc.

—— 1992. Heterogeneity of production and domestic technological capabilities in mining and mining related productive and service activities in Peru: their relevance for an environmental strategy. Lima, Peru. Research Proposal.

Núñez-Barriga, A.; Castañeda-Hurtado, I. 1994. Environmental management in a heterogeneous mining industry: the case of Peru. Report to the Mining and Environment Research Network, Bath, UK.

O'Connor, D.C. 1991a. Market based incentives. *In* Warhurst, A., ed., Environmental degradation from mining and mineral processing in developing countries: corporate responses and national policies. Science Policy Research Unit, University of Sussex, Brighton, Sussex, UK. Discussion document, Section 2, Ch. 5.

—— 1991b. The social, economic, and political context of mining sector development. *In* Warhurst, A., ed., Environmental degradation from mining and mineral processing in developing countries: corporate policies and national responses. Science Policy Research Unit, University of Sussex, Brighton, UK. Draft.

O'Connor, D.; Turnham, D. 1991. Environmental management in developing countries: an overview. Development and International Co-operation, 3(13): 75–100.

OECD (Organisation for Economic Co-operation and Development). 1991. Environmental policy: how to apply economic instruments. OECD, Paris, France.

Olórtegui, F. 1989. Diagnóstico de las Fuentes de Contaminación Atmósferica en el Perú y Bases de un Programa de Vigilancia de Calidad del Aire. Olórtegui, Pedro, OPS–OMS, DITESA.

ONERN (Oficina Nacional de Evaluación de Recursos Naturales). 1986. Perfil Ambiental del Perú. ONERN, May.

—— 1991. Problemática Ambiental del Perú. Dirección General de Medio Ambiente. ONERN, Jan. 10 pp.

OPS–OMS, Informe de Mision. 1984. Propuesta para la Prevencion y Control de al Contaminacion del Medio Ambiente Originada por la Industria Minero Metalurgica en Bolivia. La Paz, Bolivia.

Orihuela, J. 1993. Privatización de Hierro Perú. Revista Minería, No. 220, 221 (Nov 1992 – Feb 1993).

Orr, L. 1976. Incentives for innovation as the basis of effluent charge strategy. American Economic Review, 56, 441–447.

OTA (Office of Technology Assessment). 1988. Copper: technology and competitiveness. OTA, Congress of the United States, Washington, DC, USA.

Pearce, D.; Barbier, E.; Markandya, A. 1990. Sustainable development: economics and the environment in the Third World. Edward Elgar, London, UK.

Peró, M.; et al. 1992. El Medio Ambiente dentro de al Actividad Minera en la Empresa Minera "Inti Raymi" S.A. Paper presented at JICA's workshop Control de la Contaminacion Ambiental Producida por la Industria Minera. La Paz, Bolivia. Unpublished.

Peterson, S.D. n.d. RCRA's solid waste regulation and its impact on resource recovery in the minerals industry. US Bureau of Mines, Washington, DC, USA.

Pfeiffer, W.C., Fernandes, W., Guimarães, A.F., Bidone, E.D., Lacerda, D. 1989. Monitoring of mercury in the area of Carajás Project of Companhia Vale do Rio Doce — CVRD. GEDEBAM, Pará, Brazil. Mercúrio na Amazônia.

Porter, M. 1990. The competitive advantage of nations. MacMillan Press, London; Basingstoke, UK.

Portney, P. 1990. Economics and the *Clean Air Act*. Journal of Economic Perspectives, 4(4).

———— 1991a. Air pollution policy. *In* Portney, P., ed., Public policies for environmental protection. Resources for the Future, Washington, DC, USA.

———— 1991b. EPA Journal, 17(3), 37–38.

———— 1991c. The evolution of federal regulations. Public policies for environmental protection. Resources for the Future, Washington, DC, USA.

Potashnik, L. 1989. Bolivia environmental issues paper, Latin America and the Caribbean region. World Bank, Oct.

Probst, K.N.; Portney, P.R. 1992. Assigning liability for superfund cleanups: an analysis of policy options. Resources for the Future, Washington, DC, USA.

Rattner, H.; Acero, L.; Hanai, M. 1991. Mining and the environment in Brazil: the cases of gold, bauxite and tin. Workshop of the Mining and Environment Research Network, 1, 10–13 Apr, West Sussex, UK.

Reilly, W. 1989. The greening of EPA. EPA Journal, 15(4).

Reis, E.J.; Motta, R.S. 1992. O financiamento do processo de desenvolvimento. Revista Administração Pública, 1(26), 163–187.

Reis, J.A.T., Silva, M.V.T., Slabbert, T.W. 1991. Implementação de Lavra Mecanizada e Biolixiviação na São Bento Mineração [mechanized mining implementation and biox process at São Bento Mining]. In VII Simpósio Internacional do Ouro, 17–18 Sep, Rio de Janeiro, Brazil.

Ribeiro, A., Jr. 1990. Busca do Ouro cria cidade subterrânea em MG [gold searching creates "underground city" at MG). Folha de São Paulo, Brazil, 18 Nov, p. C-3.

Ribeiro, I. 1989. Visão a longo prazo orienta novos projetos [long-term foresight guides new projects]. Brasil Mineral, 7(72), 38–46.

Rieber, M. 1986. The economics of copper smelter pollution control. A transnational example. Resources Policy, (Jun).

Rothfeld, L.; Towle, S. 1989. Copper regulatory study. Environmental, health and safety regulation impacts on the southwestern US smelters and a comparison with Chile. US Bureau of Mines, Minerals Availability Field Office, Washington, DC. Unpublished report.

Rothwell, R. 1992. Industrial innovation and government environmental regulation: some lessons from the past. Technovation, 12(7), 447–458.

Ruff, F. 1970. The economic common sense of pollution. In Dorfman, R.; Dorfman, N.S., ed., Economics of the environment, selected readings. W.W. Norton and Company, New York, NY, USA; London, UK. 494 pp.

Russell, M.; Gruber, M. 1987. Risk assessment in environmental policy-making. Science, 236.

SAB (Science Advisory Board). 1990 Reducing risk: setting priorities and strategies for environmental protection. The report of the Science Advisory Board: Relative Risk Reduction Strategies Committee, SAB, Washington, DC, USA.

Salomão, E.P. 1984. The task and conditions of "garimpo" mining. In Rocha, G.A.; et al., ed., Em Busca do Ouro: Garimpos e Garimpeiros no Brasil [in search of gold: garimpos and garimpeiros in Brazil]. Editora Marco Zero, Rio de Janeiro, Brazil, pp. 35–86.

Schwarze, H.; Muñoz, G. 1991. Environmental considerations in the Chuquicamata, Potrerillos and Caletones smelters, Codelco, Chile.

SERNAGEOMIN (Servicio Nacional de Geología y Minería). 1990. Inventory of tailings dams. SERNAGEOMIN, Chile.

Silva, A.R.B. 1991. Contribuição para o Desenvolvimento de uma Economia Aurífera no Estado do Pará [contribution for the development of a gold economy in the State of Pará). Paraminérios, Pará, Brazil. 29 pp.

Silva, A.R.B.; Guimarães, G.A.; Costa, M.Q.; Souza, T.M.C. 1989. Mercury contamination in the "garimpos" of Amazônia. 3rd Brazilian Congress of Environmental Defence, 24–28 Oct, Rio de Janeiro, Brazil.

Simonis, U. 1992. Poverty, environment, and development. Intereconomics, 27, 75–85.

Skea, J, 1993. Environmental issues. In Dodgson, M.; Rothwell, R., ed., Handbook of industrial innovation. E. Elgar Press, Aldershot, London, UK. 453 pp.

Solari, J.; Lagos, G.E. 1991. Strategy for the reduction of pollutant emissions from Chilean copper smelters. In Pyrometallurgy of copper. Pergamon Press, New York, NY, USA.

Sousa, L.J. 1981. The US copper industry: problems, issues and outlook. Mineral Issues. US Bureau of Mines, Washington, DC, USA.

Stavins, R. 1983. The Tuolomne River: preservation or development? An economic assessment. Environmental Defense Fund, Berkeley, CA, USA.

——— 1987. Conversion of forested wetlands to agricultural uses. Environmental Defense Fund, Berkeley, CA, USA.

——— 1991. Toward a new era of environmental policy. In Foreman, C.T., ed. Regulating for the future. The creative balance. Center for National Policy Press, Washington, DC, USA.

Stone, J.P. 1989. Regulations of minerals industry wastes. Mineral Issues. US Bureau of Mines, Washington, DC, USA.

Suszczynski, E. F. 1986. A Questão Garimpeira [The question of garimpo mining]. Brasil Mineral, São Paulo, 35.

Sutill, K. 1993. Modernizing Porco. Engineering and Mining Journal (Jan), 32–35.

The Peru Report. 1990. Perú: the top 2500.

Tietenberg, T. 1988. Environmental and natural resource economics. Harper Collins Publishers, New York, NY, USA.

Tilton, J.E. 1990. World metal demand. Resources for the Future, Washington, DC, USA.

———— 1992. Mining waste, the polluter pays principle and us environmental policy. Colorado School of Mines, Department of Mineral Economics, Golden, CO. Working Paper 92-8, Oct.

Ugalde, A. 1992. Los Efectos Micro y los Impactos Multiplicadores Macroeconomicos de la Neuva Mineria. *In* Foro Economico 29: perspectivas de la Neuva Mineria. Instituto Latinoamericano de Investigaciones Sociales, La Paz, Bolivia.

UN (United Nations). 1988. World metal statistics year book. New York, NY, USA.

UNCED (United Nations Conference on Environment and Development). 1992. Peru: Informe Nacional [national report on the environment]. UNCED.

UNCTAD (United Nations Conference on Trade and Development). 1990. The aluminium industry of Latin America and the Caribbean: technological options and opportunities for growth. Unpublished study prepared by the UNCTAD Secretariat in cooperation with the Economic Commission for Latin America Secretariat, Geneva, Switzerland.

UNCTC (United Nations Centre on Transnational Corporations). 1985 Environmental aspects of the activities of transnational corporations: a survey. United Nations, New York, NY, USA.

UNEP. 1984. Guidelines for the environmental management of alumina production. UNEP, Paris, France. Industry and Environment Guidelines Series, Paris, France.

———— 1986a. Guidelines for the environmental management of alumina production. UNEP, Environment Programme, Moscow, Russian Federation.

———— 1986b. Multi-regional Training Workshop on Environmental Management of Aluminium Smelters in Cooperation with the Dubai Aluminium Company Ltd. Industry and Environment Workshop Proceedings Series, 8–9 Dec, Dubai City, United Arab Emirates. UNEP–IPAI.

USBM (United States Bureau of Mines). 1989. Minerals yearbook. USBM, Washington, DC, USA.

USDC (US Department of Commerce). 1979. The potential economic impact of US regulations on the US copper industry. USDC, Industry and Trade Administration, Washington, DC, USA.

———— 1988. Survey on current business. USDC, Bureau of Economic Analysis, Washington DC, USA.

USDI (US Department of the Interior). 1990. *Minerals Industry Waste Management Act of 1990*. USDI, Washington, DC, USA.

USDI–USDA (US Department of the Interior – US Department of Agriculture). 1990. Comments in response to the Environmental Protection Agency Strawman II. Sep.

Velasco, P.; Lagos, G.E. 1991. Company environmental strategies in Chilean mining, final report. Centro de Estudios del Cobre y la Minería, Santiago, Chile.

Viana, M.D.B.; Veronese, G. 1992. *Políticas* Ambientais empresariais. Revista Administração Pública, 1(26), 23–44.

Vizcarra, A.M. 1982. Tecnósfera. La Atmósfera contaminada y sus relaciones con el público. Lima, Peru.

Vogely, W. 1985. Economics of the minerals industries: a series of articles by specialists. American Institute of Mining, Metallurgical, and Petroleum Engineers, New York, NY, USA. 660 pp.

Warhurst, A. 1990. Employment and environmental implications of metals biotechnology. International Labor Organization, Geneva, Switzerland. World Employment Programme Research Working Paper, Mar, WEP 2-22/SP.207.

———— 1991a. Environmental degradation from mining and mineral processing in developing countries: corporate responses and national policies. Discussion Document for Phase I of Mining and Environment Research Network, 10–13 Apr, Wiston House, Sussex, UK.

———— 1991b. Environmental management in mining and mineral processing: challenges for sustainable development: objectives, methods, programming and output of the Collaborative Research Network. Science Policy Research Unit, University of Sussex, Brighton, Sussex, UK.

———— 1991c. Metals biotechnology for developing countries and case studies from the andean group, Chile and Canada. Resources Policy (Mar), 54–68.

———— 1991d. Technology transfer and the development of China's offshore oil industry. World Development, 19(8), 1055–1073.

———— 1992a. Environmental management. *In* Mining and the environment: the Berlin guidelines. Mining Journal Books, London, UK. Chap. 7.

——— 1992b. Environmental management in mining and mineral processing in developing countries. Natural Resources Forum, Feb.

——— 1992c. The limitations of environmental regulation: an argument for technology policy to promote environmental management in mining. Paper prepared for the 1992 John M. Olin Distinguished Lectureship Series in Mineral Economics, Nov. Colorado School of Mines, Golden, CO, USA. 34 pp.

——— 1992d. The political economy of mining and environment: the policy challenge of environmental management in the Southern Africa Development Community. *In* Bergstrom; Zulu, ed., Proceedings of the Workshop on Mining and Environment in the SADC Region, 1–3 Dec, Lusaka, Zambia. Southern Africa Development Community, Lusaka, Zambia.

——— 1993. Environmental regulation, innovation, and sustainable development. Paper presented at the 3rd International Workshop of the Mining and Environment Research Network, 14–17 Sep, Wiston House, West Sussex, UK.

——— 1994. Environmental degradation from mining and mineral processing in developing countries: corporate policies and national responses.Organization for Economic Co-operation and Development, Development Centre, Paris, France. Development Centre Document.

Warhurst, A.; MacDonnell, L.J. 1992. Environmental regulations. *In* Mining and the environment: the Berlin guidelines. Mining Journal Books, London, UK. Chap. 2.

WCED (World Commission on Environment and Development). 1987. Our common future. Oxford University Press, Oxford, UK.

Webb, R.; Fernández, B.G., ed., 1991, 1992, 1993. Cuanto, Perú en Números: Anuario Estadístico. Prepared by.

Winters, R.; Marshall, L. 1991. Where's the recovery in RCRA: the re-mining of non-coal abandoned mine sites. Proceedings, 12th Annual Meeting of National Association of Abandoned Mine Land Programs, 9–10 Sep.

Womack, J.; Jones, D.T.; Roos, D. 1990. The machine that changed the world. Macmillan, New York, NY, USA.

About the Institution

The International Development Research Centre (IDRC) is committed to building a sustainable and equitable world. IDRC funds developing-world researchers, thus enabling the people of the South to find their own solutions to their own problems. IDRC also maintains information networks and forges linkages that allow Canadians and their developing-world partners to benefit equally from a global sharing of knowledge. Through its actions, IDRC is helping others to help themselves.

About the Publisher

IDRC Books publishes research results and scholarly studies on global and regional issues related to sustainable and equitable development. As a specialist in development literature, IDRC Books contributes to the body of knowledge on these issues to further the cause of global understanding and equity. IDRC publications are sold through its head office in Ottawa, Canada, as well as by IDRC's agents and distributors around the world. The full catalogue is available at http://www.idrc.ca/books/index.html.